Building Procurement

Second Edition

Alan Turner

MACMILLAN

690.0684 TUR

First edition 1990
Reprinted 1994
Second edition 1997

Published by
MACMILLAN PRESS LTD
Houndmills, Basingstoke, Hampshire RG21 6XS
and London
Companies and representatives
throughout the world

ISBN 0–333–68809–0

A catalogue record for this book is available
from the British Library.

This book is printed on paper suitable for recycling
and made from fully managed and sustained forest sources.

10   9   8   7   6   5   4   3   2
06   05   04   03   02   01   00   99

Printed in Hong Kong

Contents

List of Diagrams

client advisers and the like, it is understandable that people from many different backgrounds now want to be 'the client's representative', to be the principal person.

It will be interesting to see if building procurement continues to come further towards the centre front of the stage – if successfully done it is an art, and many new players are learning that art.

Preface to the Second Edition

Writing the first edition of a book is one thing, revising for a second edition is quite another; likewise with a preface.

On reading the preface to the first edition of this book it still stands very adequately as an introduction. It sets the context of writing about building procurement in 1990 and much of it is just as valid today. This preface aims to give the changed context of 1996 in which building procurement is now carried out.

In the six years since 1990 the construction industry has been affected by a number of major changes. First, the industry entered, around 1989/90, its most prolonged recession since the Second World War. Of course in 1990 this was not apparent and only in retrospect can the true severity, depth and length of the recession be seen. Recovery has still not fully occurred, at least to match the somewhat unreal levels that existed when I wrote the first edition in 1990. Output, measured by most of the usual statistics, shows supply continuing to exceed demand, at least relative to the boom of the 1980s which reached its peak in 1989. In true terms tender prices are barely higher now than they were in 1989. Of course costs have risen but the end price to a 'building procurer' has rarely been such good value for money as it is today.

The consequences of the recession for clients, consultants and contractors have been varied. Clients have enjoyed a most competitive market since 1990, with too many organisations offering to supply them with building products and services, often at less than cost. The consequences of recession for consultants and contractors have been drastic cost-cutting, reduction of staff numbers to match falling workloads, attempts at improved efficiency and so on. During all of this, failures, liquidations, voluntary administration arrangements, enforced mergers, takeovers and other measures have been the norm for many an organisation. It is estimated that around 400 000 people have left the industry since the peak of 1990 with an estimate, as I write this, that perhaps another 30 000 may go in the next twelve months. Polarisation of companies and partnerships into large or small units is now taking place. Growth by acquisition is common and many old names, particularly among consultancies, have disappeared into general companies that offer practically any service, not only in the construction industry.

Change in government attitudes has led to more and more privatisation, 'opted-out' schools, health trusts and therefore more and more

potential procurement organisations, decentralisation of services leading to more 'amateurs' becoming responsible for procuring buildings.
The central client role of government is less present today but government is more interested in competition, compulsory competitive tendering, value for money and in measuring performance by 'benchmarking'.

Global markets and international companies seem to have become
more intrusive in the UK market. Foreign ownership and participation
in UK construction companies and consultancies is now gaining pace,
invariably bringing a change of attitudes and practices. 'Partnering' by
clients with consultants and contractors is fashionable. The government
has retained its very tight grip on public expenditure and has introduced the 'Private Finance Initiative' in an attempt to have private organisations provide public facilities for the government to lease or,
sometimes, buy.

Professional institutions and trade associations in the industry have
faced a relative drop in their membership and therefore in their in-
comes and are increasingly reluctant and/or unable to support other
expenditure on industry initiatives, in addition to their membership of
more 'pan-industry' bodies, such as the Construction Industry Council,
the Construction Industry Employers Confederation and the Constructors Liaison Group.

In the middle of the recession a major review of the construction
industry has been carried out by Sir Michael Latham, resulting in the
publication of two reports, an interim report, *Trust and Money* in 1993
and the final report, *Constructing the Team* in 1994. These reports
were generally well-received by the industry and caused, among other
things, a series of 'working groups' to be established to develop such
subjects as client briefing and codes of practice for clients of the construction industry. Other subjects examined or still in process are qual-
ity/price mechanisms for selection of consultants, standard qualification
documents for public sector work, training, the image of the industry,
equal opportunities, education of construction professionals, liability
and latent defects insurance, partnering and a 30 per cent reduction in
real construction costs. This last point, apparently asked for by major
clients of the industry, is to be achieved by the year 2000. At the time
of finalising the second edition of this book, in mid-1996, it is fair to
say that much has been promised by setting up the 'working groups'
and some of the reports now being published make interesting reading.
It remains to be seen how much can actually be delivered by the industry at large following such a focus of attention on its practices.

It is crucial to realise that the client has become more and more
important, quite understandably, since the late 1980s in the process of
building procurement. Now no self-respecting construction body can
speak without having visibly taken counsel with a client body. The

Construction Clients' Forum (CCF), the Business Round Table and the Construction Round Table (CRT) have been formed, the latter 'committed to making significant improvements in the performance of the industry'. The CRT is a small group of leading customers of construction from different market sectors. Collectively it invests over £3 billion a year in construction (government and its agencies probably invests around £5 billion annually). Whilst this is laudable, and the aim of these bodies is to bring the best of 'good practice' obtained by them from the industry to more and more occasional clients, building procurement remains for far too many an unsure, uncertain hazard.

The Construction Industry Board (CIB) has been set up following Latham and the industry now has the most structured, formalised machinery ever, with, it is claimed, much better access to the government. All this has come about following complaints of the industry (clients, contractors, subcontractors, consultants, suppliers – not apparently directly trade unions) and particularly some of the industry's foremost, therefore big, clients.

So it is hoped that something more cogent, long-lasting and fundamental will come out of Latham than has generally been discernible following former major industry reports, such as *Action on Banwell*, 1967, *Banwell*, 1964, *Emmerson*, 1962, *Phillips*, 1950, and *Simon*, 1944.

Another part of the findings of Latham was that the industry was too confrontational and that procedures, contracts and dispute resolution methods such as adjudication, aimed at reducing conflict and dealing with disputes more quickly, should be introduced. The right to adjudication has been made statutory and it remains to be seen if the industry finds this leads to less litigation or arbitration. Fairer construction contracts have been proposed by Latham and the CRT has announced a number of major clients and construction companies that have committed themselves to observing the twelve principles of fair construction contracts, as seen by CRT. The Chartered Institute of Purchasing and Supply (CIPS) keeps a register for inspection of organisations committed to fair construction contracts.

In the area of building procurement the role of a 'principal adviser' is needed more and more to give independent advice at crucial times to clients, particularly before any early decisions and appointments of consultants and contractors are made. The advice of a principal adviser is particularly of value, and is usually crucial, to those who have never procured a building before. It is hoped that education, training and promotion throughout the industry will make the role of 'principal adviser' more widely understood by clients and by the industry.

ALAN TURNER

Acknowledgements

Thanks are due to the Business Round Table for permission to reproduce pages 6 and 7 from their publication *Thinking about Building*, published by the Business Round Table in 1995. This publication, updated by the Construction Round Table following the advice published in the National Economic Development Office (NEDO) publication of 1985, remains the clearest layman's guide for anyone thinking of building procurement. In particular I thank Eur.Ing. Henri Pageot, BSc, CEng, FICE, the Executive Director of the Construction Round Table (CRT), for ideas on a number of construction industry issues exchanged during discussions in his office and for access to a number of CRT reports published during 1995 and 1996.

My direct knowledge of construction management, unlike management contracting, has so far been limited and therefore I sought the view of two organisations that would be representative of experienced clients and of experienced contractors. First I thank Neil Smith, of Laing Group Contractual Services, for his help, and particularly Alan Page of Laing Management Limited, for a very full explanation of construction management (and management contracting) as provided by John Laing and for reference to the University of Reading report, published in 1991, after the construction management forum. Second, I record my thanks to Christopher Strickland, development director, and Julian Vickery of Greycoat plc for discussions on their approach to construction management.

Since the first edition I have learnt some of the mysteries of word-processing and have carried out production of most of the new text myself. In those cases where I have not, I thank my previous secretary, Rena Monks, for again being on hand to help out. Thanks are due to my daughter Siân for assistance in preparation of the illustrations and diagrams in the book. I also thank Professor Roy Morledge of The Nottingham Trent University for a sight of a number of drafts of *The Procurement Guide* to be published by the Royal Institution of Chartered Surveyors in 1996. Last I thank my brother, Dennis Turner, BA(Hons), FRICS, MCIOB, ACIArb, who has written several books on building contracts and building contract disputes, for ideas that we have exchanged during a number of discussions on this and other books.

Since 1994 I have been in practice on my own account as an arbitrator and quantity surveyor, principally giving advice on procurement and contract disputes. I thank the clients I have had in these areas for

confirming two views. First, that with the best procedures and contracts but with the wrong attitudes in a team conflicts may still arise.
Second, that poor briefing, poor documentation and poor administration can invariably be overcome and a project made to work, albeit
with stresses and extra costs, if attitudes of goodwill exist on all sides.
But even with goodwill the worst parts of the construction industry, its
clients, consultants and contractors, remain a wretched advertisement
for mismanagement and are therefore too often the public image of
'building'.

ALAN TURNER

Part 1

1 Reasons for Building

Three fundamentals

Since writing the first edition of *Building Procurement*, as the result of advising clients and conducting seminars on building procurement, I have condensed the essentials of 'procurement' into three fundamentals. These essential, fundamental considerations apply to the purchase of any product or service by any individual or by any organisation.

Whenever we consider making a new purchase we instinctively go through the following three questions, in the following order, before deciding to buy:

1. HOW GOOD?
2. HOW SOON?
3. HOW MUCH?

This routine applies to the purchase of billions of items by billions of customers or clients every day. Building procurement, purchase of buildings, is no exception.

Some may query the order of these questions but I am convinced that the priority of consideration of these fundamentals is always 'how good', 'how soon', 'how much'.

If a product or service is not perceived as being of any use by a potential customer, that is, 'how good is it for me?' is no good, then that customer, quite logically, does not want to know how soon he or she can enjoy the product/service or how much it will cost. 'How good', that is, perceived quality, is the starting point in a customer's search for any product. If a product/service is not perceived as being of any use, the fact that it is available ('how soon') and/or that it costs next to nothing ('how much') has virtually no interest. Promotion of most sophisticated products usually extols the qualities of how good the product is and only then may go on to state how soon the product may be bought and/or the price of that product. Of course customers ask the other two questions quite soon after asking 'how good'.

Because the pace of life in developed countries is still aimed at shortening the period between 'a wish' and 'fulfilment of that wish', questions of 'how soon' quickly follow and then 'how much' is usually not far behind. Buildings are no exception to this rule but in the vast majority of cases they cannot be 'stockpiled' as can many other products.

Buildings cannot generally be produced in advance, to call off a production line, at least not such that they will be in the location and in the form that most clients will want them, and this makes questions of 'how good', 'how soon', and 'how much' even more difficult to answer in building procurement.

Unfortunately, comparisons between products of the automotive industry and those of the construction industry are often made, generally too often made. I will return to this point later in more detail. A comparison that is validly made is one of quality, of 'how good'. New cars are sold on the perception of quality, 'how good' – and far too seldom can buildings be sold as easily in relation to their quality. The perception of the construction industry by too many construction clients remains one of poor or indifferent quality. In my discussions with the Construction Round Table, I was impressed by their aims and efforts to improve the general quality level of the industry, so that the average of the industry approaches much closer to that of its best, because its best is excellent, both in concept and in execution. The CRT are 'committed to making significant improvements in the performance of the industry' (this quotation is taken from the preface in all their publications) and hopefully this will come about.

Because in everyday life many of our purchases are relatively trivial the even more fundamental question than the three 'hows', as discussed above, is often not asked. That is the question 'Why?'. Why do we want a product in the first place, what will it do to satisfy our needs?

Why?

An essential question to ask before embarking on any exercise or enterprise beyond that of the transitory or trivial is 'Why?'

An exercise, that is, a set of movements, tasks and so on designed to train, improve or test abilities, may perhaps be transitory or trivial, especially to those not involved. But an enterprise, that is, a project or undertaking, especially one requiring boldness or effort, is generally not started as a transitory or trivial matter.

Construction is undeniably such an enterprise, an act of boldness even for the simplest building. For modern, complex buildings it involves the commissioning, management, design and assembly of huge amounts of raw materials and the use of considerable labour resources over a long period of time – so why is it carried out? In simple terms construction is only necessary when no other building exists or appears to exist that either meets or appears likely to meet the needs of a client. It is self-evident that the client is the only reason for building and is the key to why and then to how a building is obtained, how it is procured.

In the current version of *Thinking about Building* some basic questions are asked of a client in relation to its business. Where is the business going? Should it:

- stay where it is and improve its existing workspace?
- relocate to new accommodation?
- if so, lease or buy?
- acquire a new building 'off the shelf', or one tailored to the company's specific requirements?

A building is a key contributor to a business operating efficiently and effectively. Energy considerations, other running costs, morale of the occupants, image of the company as seen by the staff and by the customers, may all be affected by the building(s) used by a business.

Options for consideration can sometimes seem bewildering. At some point a prospective client will almost certainly need some expert advice. But before asking for outside advice a client should:

- set out the in-house skills that it has for going about 'building'. It may have experienced the process before or it may be a 'novice'. The skills may be many or few and it is best to be truthful on this vital point
- look for guidance from those who understand the business of the company. These may include a bank manager, the company's accountant, lawyer, property surveyor and other business colleagues
- ask who in the company will represent the client. Which person of sufficient seniority is able and willing to act as a single point of contact throughout the total briefing, planning, design and construction process?
- decide if there is such a person and whether or not he (or she) is available for the duration and with the intensity required
- decide with or without such a person how the business will manage the process of building procurement
- then decide what other expert advice the business will need, from an independent 'principal adviser', before any other consultants and/or contractors become involved.

Clients

As stated in the preface to this revised edition a major review of the construction industry was carried out by Sir Michael Latham in 1992/93 and his findings were published in July 1994 in a final report, called *Constructing the Team*.

In that report the central and crucial importance of the client in building procurement was emphasised by Latham many times over. That the customer is king is very clear. Whilst this is self-evident it had been overlooked and that was a sad commentary on the state of the construction industry in the mid-1990s. *Thinking about Building*, as revised by the CRT, and the efforts of working groups established to implement the Latham Report will hopefully keep this point in the forefront of actions of all suppliers of products and services to construction clients.

Some of the Latham 'working groups', set up in late 1994, were briefed to bring clients and the industry closer together. It is disappointing that the industry apparently needed, yet again, to review and produce guidance on some fundamental topics. At the time of drafting this revised edition a report called 'Constructing success,' (to assist clients in acquiring their desired end product and obtaining value for money in so doing) will be published towards the end of 1996. Very worthy work is being carried out in these areas so that benefits may accrue to project teams and to their clients during briefing.

Part of the effort to improve the construction industry's performance has been to invite 'clients' into the industry bodies, old and new, so that the influence of clients can be directly contributed. To this end the Construction Clients' Forum (CCF) was formed. During my period as chairman of the National Joint Consultative Committee for Building (NJCC) invitations were issued to the CCF to become observers and eventually participate in the work of formulating 'good practice' for the building industry. The CCF is a member of the Construction Industry Board (CIB), set up in 1995 to implement the Latham recommendations. The CCF has found it difficult to attract a wide, representative membership. At the time of writing (1996), it has not been seen by many as having been able to make as great a contribution as hoped for by the construction industry. Membership of the CCF is mainly from property companies, major retailers and other clients that regularly build. This may be satisfactory in the short term. For whatever reason it has not been able to obtain membership from local authorities, from a good cross-section of 'quangos' or from representatives of those organisations that build very infrequently. Perhaps this cannot be solved. Does an organisation that builds only once or twice, by definition, wish to join a collective organisation that is mainly comprised of other client organisations that build regularly?

The essential point to emphasise is that clients come in many shapes and sizes. Whilst the categories described over the following pages embrace, broadly, the main client groupings, a very large proportion of the industry's clients only build once or maybe twice in a lifetime. Advice in books such as this, in *Thinking about Building*, in reports of post-Latham working groups and so on have an aim to help clients unfamiliar with the industry. An additional aim is to make the relatively poor performance of the average industry organisation improve so that it becomes much closer to the 'best practice' available from the industry's leading organisations in many, many more instances. Part of the effort is to make clients themselves more aware of what the industry can provide and, perhaps as importantly, the industry aware of the objectives of their clients. If this is attained the effort will have been worthwhile.

Building clients can be categorised in several ways. Individuals, groups or partnerships of people, corporate bodies; private and public clients; clients who build once or rarely; those who build often; those who build for owner occupation; those who build for investment or as developers; those who act as agents or agencies for those who will eventually occupy the building; those clients who are combinations of the above.

Clients, customers for building, once they become involved in building contracts and/or professional services' conditions of engagement, are often termed 'employers' but for this section of the book they remain as 'clients'. It is useful now to look at the overall objectives of different types of clients and how those clients are part of the property scene.

Objectives of building clients

In simple terms the purpose of any building development is to provide accommodation for occupation either for the person carrying out the development or for someone else.

Reference has already been made in the preface to customers or clients. Construction clients are not uniform or average organisations. It is dangerous to generalise because the objectives of one client may be quite different from those of another. Therefore it is useful to consider clients in the following five categories, principally because the needs of different clients lead to different priorities and put different demands on the construction industry.

The objectives of clients vary according to the nature of the client and those objectives may well influence quite strongly which of the building procurement paths should be chosen.

For the purpose of this chapter clients can be divided into five categories:

1. Property and development companies
2. Investors
3. Occupiers
4. Local and central government authorities
5. Quangos

Property and development companies

The prime objective of a property company is to make a direct financial profit from the process of development. In this respect it is no different from any other company making a product for sale. It brings together services and raw materials and then processes them into a product that is sold in the market-place. The ultimate customer for the product may be known beforehand, as when a building is prelet, or may be entirely unknown. A property company normally produces and sells direct to the customer without a 'middle man', but often takes advice from or uses property agents.

Within the overall objective of profit, however, the aims and activities of property companies can vary. Some companies specialise in terms of location, maybe as a local company only developing in and around a particular town or area it knows well. Other companies specialise in certain types of property, for instance, industrial development, offices or shopping. Some specialise in a particular kind of development process such as altering or improving existing buildings, known sometimes by that ugly word 'refurbishment'.

Almost any form of specialisation should enable companies to acquire above-average knowledge and experience of a particular market or of a particular location, so making it easier for them to succeed where others may fail.

Alternatively, some companies deliberately avoid specialisation and try to spread their risks, either geographically or in terms of types of property. They aim for a balance within their development programme between, for example, offices, industrial premises and shops or between one town or region and another.

Financial objectives of property companies also vary. Some companies seek to sell each project before it is completed or during the design and construction period, taking any immediate capital profit. Others try to retain an interest in the property and look for income related to retaining the property so enabling them to enjoy future growth in rental income. There are no fixed rules.

Investors

As with property companies the prime objective of investors in property is direct financial gain but investors tend to take a longer view than property companies and are usually more concerned with the flow of income over an extended period of time. When they do become involved in property development itself, they often do so as a means of producing an asset. The holding of the property asset will then hopefully produce an income that will give them an adequate return on the capital invested and an opportunity to see that both income and also the capital investment grow. In order to be persuaded to take on the risks associated with development, rather than simply acquiring a property when it has been completed and let, investors usually need a higher overall return. Because of the combination of possible investment and development gains they may often be prepared to accept a lower immediate return from the development process than would a property developer.

Naturally policies of investment companies differ but they all tend to seek a balanced portfolio of property uses, rather than specialising in one particular use. Most companies try to spread their investments geographically although some may concentrate on the area or market sector they know and can manage best. Investors tend to be cautious and dislike unconventional buildings and uses, unconventional lease terms, and perhaps (to them) less conventional building procurement methods. They may avoid properties that will involve substantial management and often prefer, for instance in the case of offices, a building to be let to a single or at least few tenants.

Occupiers

An occupier tries to obtain a building that will be best suited to his particular needs. These may be relatively simple or may be very specialised as, for example, where a commercial or industrial process is involved. An occupier's prime objective is to have a building within which he can best carry on his business. Because of this the building that the occupier seeks may on occasion have little or no general market value.

Naturally it is planned that profit will arise from the use and benefits derived from occupation of the building but it may be that the open market value of the completed building is less than the total cost of development – a case of cost exceeding value. Sometimes the prospective occupier may need to compromise and modify his plans so producing a more standard type of building, because such a building should have a more established market value so permitting greater flexibility in the event of its disposal, or the need to raise a loan on the security of the building.

Commercial and industrial companies often become involved in schemes to develop for their own occupation either as extensions to their existing premises, or on new sites. The efficient control of the process of development should be just as important to them as it is to a property company or to an investor involved in development. Even though one of the main elements of development risk, namely the finding of an occupier, is absent, most of the other factors of risk remain, principally those of possible variations in price and in the programme for completion of the development.

In some ways the risks may be greater for an occupier. An industrialist or retailer may decide to expand his accommodation in order to house a new process or respond with a new marketing approach. The cost of the building may be only a small part of the total project costs whereas a delay of two or three months in construction completion may delay a new production line or a new service and lead to loss of orders, to missed opportunities, and then to a ruined reputation. Therefore, it is important that occupiers carrying out their own development should understand:

- the time that needs to be taken in arranging for and carrying out projects
- the sequence of events in the development process
- the range of possible variations in building costs and the alternative methods used to control development expenditure.

Chapter 2 emphasises the importance of a 'principal adviser' to an occupier seeking to develop his own premises.

Local and central government authorities

Local and central government authorities are often now involved with building projects and sometimes with property development for profit – they can be both occupiers and developers. As occupiers of buildings, local and central government authorities often build regularly and should understand well the development and building process. They may also be involved in some way directly for profit or by sharing indirectly in profit that arises from development.

When local authorities are building clients, acting as property developers, there are probably two further considerations present that are not usually so with 'private' property companies. These are public accountability and the need for an authority to consider the overall desires of its community and any possible effects, other than monetary, of the development/building proposals on the environment. More sophisticated techniques of project evaluation, such as cost–benefit analysis, may be required in these circumstances whereas they would not normally be required in development appraisal for a private client.

Quangos

Since the publication of the first edition in 1990 many more quangos – quasi-autonomous non-governmental organisations – have come into being. Health trusts have become responsible for administering National Health Service provisions at local level. Independent 'opted-out' schools have increased in number and can procure buildings without reference to their local education authority. The overall effect of this has been that authorities that had previously been subject to more centralised controls, often with procedures that had arisen over a long period of dealing with the health service at national level or with county or other local authorities, have disappeared or at least lost many of their procurement activities. Procurement often now takes place with more 'facilities officers' and external clients' representatives becoming involved. The objectives of these bodies include satisfying 'good value' criteria and remaining accountable to local governors and managers. These constraints not only put more control into smaller units but have often introduced more diverse procurement procedures. Experienced officials, familiar with the construction industry, are not always so obviously in key positions in building procurement in some of the quangos and a criticism of some of these newly formed bodies is that, as they often do not have experienced staff, then 'amateurs' are sometimes responsible for procurement of buildings.

In my submissions to the Latham review, when I was chairman of the National Joint Consultative Committee for Building (NJCC) in 1994, I wrote as follows (reprinted from page 3 of the Latham Report, *Constructing the Team*, 1994):

> Nowadays the NJCC has no means of ensuring that all housing associations, trust hospitals, grant maintained schools, private government agencies, utilities companies, etc., are aware of the best current practice and changes in the construction industry . . . We also notice individual Government Departments operating different procurement practices and that this has become more pronounced since the demise of the PSA. Unless an effective communication network is established, as the industry is called upon to play its part in the economic recovery, more and more cases of bad practice will come to light and the thrust of the Interim Report [that is Latham's *Trust and Money* report in 1993] may be of no avail.

'Quangos' may be principally interested in building procurement as owner-occupiers but may be tenants or become developers. In each role their requirements may change and it is important to establish the reasons and constraints that each has in seeking to build.

Property

It is necessary to put *building* into a wider context, into that of *property*, and then to come back and consider building itself. Put simply, as an economic category buildings exist as property. The land on which buildings stand is generally the most critical element in property and the permitted use of that land, allowed by any planning regulations, is the single most important factor. If the most exciting, commercially viable scheme conceivable cannot be erected because it falls, for instance, within the Green Belt, that (at present anyway) is that.

Property and the valuation of property needs to be understood in simple terms. Because property, that is, land and most often the buildings on it, is viewed in our economy as long-term stock it has a long-term investment value, unlike items such as cars, consumer goods, food, holidays and services generally. It is important to understand that property has a value and its value often changes according to factors such as its location, the type and current use and attractiveness of any buildings on the land, any future change of use which may be desirable or allowable for the land according to any planning rules of the particular country and so on.

For example in Britain the development of London Docklands has taken place relatively quickly because planning legislation, combined with tax and grant incentives, has been relaxed to encourage developers to provide new buildings in an area that at the beginning of the 1980s was a prime example of urban decay. The valuation of the buildings erected in Docklands has followed broadly the usual criteria for commercial development – that is, a development appraisal with benefits that exceed the total development costs has meant it has been worthwhile erecting new offices, shops, light industrial buildings, houses and so on. The state of demand and the equation of development value exceeding costs may or may not continue on an even course. Also opinions on the overall result and success of Docklands will very probably continue to differ quite widely.

Generally speaking all new construction or modification of existing construction will take place only if a development appraisal makes sense. This has come increasingly to be the case not only for private clients but for the public sector also. Exceptions to this have been buildings for defence, law and order, health and education but development and investment appraisal techniques are used in some instances even here.

As stated earlier buildings are generally erected for one of two reasons – first for investment, and long- or short-term holding by a property company, pension fund or whatever, or second by an organisation that wishes to use the building, namely by an occupier. Whereas a building can be erected almost anywhere, subject to certain physical restraints of soil or climate, the construction of a building is much more likely to be constrained by the time-honoured rules of property. Because property development is by its nature uncertain it involves quite an element of risk and before any building is commissioned, generally before any land is purchased, an appraisal process of the proposed development will take place – a developmental appraisal.

It is still probably true that the first three rules of successful property development are location, location and location. Quality of concept and of design have however now become very much more important and location may not take the first three places quite so easily as it once did. The 'three rules' saying draws attention to the fixed nature of buildings and therefore the extreme care required in deciding where to put a shop, a shopping centre, a new automobile plant and so on. Some property is more sensitive to location than others, for instance a speciality shop or restaurant, whereas a warehouse may satisfy locational criteria provided it is near to a major road, rail junction or whatever.

If the first three 'location' rules may not have changed, location itself is of course changed over and over by time – areas can become more desirable in society's expectations and conversely they can also decline. Time is also a critical ingredient in development because it is necessary to ensure that a development is produced at a time when there is a demand for it.

It is now appropriate to understand a little of the development process before looking at 'building' itself, in Chapter 2.

Development process

The development process can be seen in five phases:

1. **Development appraisal** or evaluation of the project generally.
2. **Development preparation** for the project to proceed to completion.
3. **Development implementation** of management, design and construction of building – building procurement.
4. **Disposal** of the newly created space by occupation by a tenant or occupier, perhaps sale of the property (if this has not occurred between phases 1 and 2 or 2 and 3 above).
5. **Maintenance** of the project until renewal occurs.

Developmental appraisal

Developmental appraisal or evaluation is the first phase and the most critical – errors made here probably spell success or failure irrespective

of the later phases of development unless fortuitous markets, dispro-
portionate rates of inflation between rents and building costs and so
on produce windfall solutions (or escapes) for the client. The criticality
of this phase will be principally for a client whose business is property
investment or property development but also sometimes, perhaps to a
lesser extent or with a delayed reaction, for a client who will become
the occupier of the building.

Development appraisal is a subject on its own but it is important to
see building construction and its procurement within the context of the
development equation. Development appraisal will have the following
constituents.

(a) Market research

If the building is for investment it is essential to establish that there is
likely to be a market for it in the location, at the time it will be produced, and at the price it will have cost – an overpriced, late-com-
pleted office building in the wrong location will have failed. Similarly
if the building is for owner occupation it is equally sensible to evaluate that its cost will produce profit for its occupier at the location,
time and price it will have cost – a late factory, or office, constructed
over budget, in the wrong location naturally will be more than a disappointment.

Property research made a rapid move from its academic base into
the market-place during the second half of the 1980s. Major property
agents need to be able to supply their clients with research data, as
opposed to only judgment, on the prospects for each new development. Technological revolution alters the kind of buildings that are
required and hastens their obsolescence. Buildings of the late 1950s
and 1960s have generally proved unsuitable for new office and retailing environment. It is particularly important for today's property developers to try and look far enough into the future to avoid immediate
irrelevance of their product or fairly short-term obsolescence.

In addition to this, property itself is no longer seen, as in the past, as
an essential component in any balanced investment portfolio. Rather,
property is seen today as only one of a number of investment markets
and it has to compete for its share of funding, along with equities,
fixed-interest stocks and so on in the UK and in other global markets.
The sheer size of some of today's schemes means that international
funding is involved.

Market research now regularly involves sophisticated location studies,
surveys of a district, of the growth characteristics of a town, of neighbouring towns, of populations, of economic and income groupings and
so on. If this seems a great deal of research it should perhaps be seen
against other investment decisions. For instance no industrial manufac-

turer, in a fast-moving and complex market, would put £10 million into a new product without extensive research and property developers may not be able to continue, on occasion, to gamble similar sums without good market knowledge.

The results from market research may heavily influence the building procurement process as, for instance, when a developer waits for the appropriate moment and then, after seemingly inordinate delays, wants to commence building quickly with apparent undue haste and undue concern for the working methods, procedures and desires of his building professionals and contractors.

(b) Financial evaluation
Development appraisal usually employs one or more of three methods for the financial evaluation involved, once market research has established a need for development.

The *conventional technique* compares total expected costs with total expected revenue to discover whether or not the project will produce an adequate return, in terms of either a trading profit or of an investment yield.

The main development variables are:

- land price
- building cost and gross area
- rents and net lettable area
- interest rates, short-term borrowing costs
- investment yields
- programme for development
- purchase and sale costs.

More sophisticated techniques can be adopted by using the *cash flow method* which takes account of the estimated timing of expenditure and revenue more accurately than the *conventional method*, over the programme of the development. A further refinement in appraisal techniques has been the introduction of *sensitivity analysis* into the appraisal equations. Naturally sensitivity analysis confirms that the most critical items are the investment yield required, the rent expected and the estimated construction and associated costs. Generally the investment yield required is not likely to change during the timescale of a development whereas expected rentals and/or construction costs may very well change. A change of plus or minus 10 per cent in yield, rents or building costs will usually have significant effects on the expected success or failure of a project.

(c) Investment appraisal

Whereas *market research* and *financial evaluation* described earlier are vital parts of a project appraisal it may be that the project needs to be appraised as an investment in a strategic contest. *Investment appraisal* is a systematic approach to presenting information to assist in decision-making.

It is a term currently used to cover a range of evaluation techniques used when making investment decisions and it has become popular within an overall process which is used particularly in local and central government, often referred to as capital budgeting or capital planning. As such, investment appraisal is not always concerned only with financial factors but with the evaluation of both financial and social benefits and costs. Sometimes these social costs and benefits can be measured and valued in financial terms but, when this is not possible, a system of ranking may be used to indicate and compare non-quantifiable factors.

The Appraisal of Capital Investment in Property – A Guide for the Quantity Surveyor, 1989, the Royal Institution of Chartered Surveyors, gives a general framework and suggestions for the practice of investment appraisal and is a useful introduction to the techniques and considerations involved.

Development preparation

Although the development process may be iterative and/or it may need to proceed simultaneously with a number of sites, at some point a decision to proceed only with one site, after initial development appraisal(s), will eventually be taken.

A client will then need to try to quantify more precisely the broad items used at appraisal stage. As this is done the initial evaluation most probably will need to be revised and, if the figures then show an unacceptable return, the project may have to be abandoned.

For this reason a developer will attempt to delay any substantial commitment, such as the purchase of land, until the greater part of the initial evaluation has been thoroughly tested. It may well be, however, that the land has been purchased before all the factors are known, and if so, any evaluation should allow for the element of risk that is thereby undertaken.

During the period of development preparation the following, or a substantial part of it, may need to be completed:

- study of the physical extent and nature of the site
- investigation of the extent and nature of any interest(s) in the land that is being purchased

- preparation of more detailed outline drawings
- submission of a planning application(s)
- negotiations with the planning authority
- obtaining planning permission
- obtaining short- and/or long-term finance in commercial projects, perhaps securing a preletting of a whole or part of the project
- in all cases, more thorough investigation of the market and establishment of levels of price or rent
- preparation of more detailed estimates of construction cost and fees, some discussion of the procurement route likely to be selected and possibly some preliminary discussions with contractors and/or construction advisers.

When the development preparatory work has been completed, the project may need to be evaluated again. This may be because of a significant period since the last evaluation or because some significant change in national, local or market circumstances may call into question the original appraisal. Ideally it is important for a client to refrain from committing himself to as many costs as possible until he is satisfied that his initial evaluation is correct and/or that any revised evaluation confirms a viable scheme. Until land is acquired, costs to the client are unlikely to be substantial but thereafter a client is committed not only to a land purchase price but also to a particular location. Many clients will only want to commit themselves to land or property purchase subject to their obtaining the appropriate planning approval and it should be remembered that negotiations with a planning authority can take up a considerable proportion of the total development preparation time.

Development implementation

The implementation of the development phase brings together the financial, design and construction phases of the project. By this stage a commitment has been made to a particular site, to a particular building, hopefully at a particular price and at a particular time. Once a decision is taken to proceed into the development implementation phase it is most unlikely that changes of any significance in the scope and intent of the works can successfully be made, once a procurement route is chosen and implemented. However any of the procurement routes may have the possibility of unwanted cost and delay and for this reason the client will need to remain closely involved or to appoint an adviser to do so on his behalf.

The options of building procurement and choosing from them the appropriate route are detailed in Chapters 4 and 5.

Disposal

At the completion of building, the planning, management, estimation and construction on the project will all have become reality. The building has been designed and built and it should be occupied either by the owner for whom it was designed or by a tenant. The building may then be sold or very often it may have been sold already before building completion by its developer, by its original owner, or whoever. At this stage of the development process, once the building has been completed, it physically becomes the property asset, the property investment that was planned and it can be sold, mortgaged, used as a security and so on probably several times throughout its life.

Maintenance

The building will need to be maintained during its life. This will be carried out by its owner, by its tenant or by a combination of the two either directly or by the use of property agents/maintenance companies. Property maintenance is a specialised subject and the responsibilities for maintenance will vary. An owner-occupier will have the maintenance responsibility and, if the property is of any significance, will probably appoint maintenance surveyors to advise and to implement regular, planned maintenance to ensure that the capital value of the property is kept in balance with its economic value to the owner.

A tenant of offices or shops may well be under an obligation of a full repairing lease and will therefore have the responsibility to maintain to agreed standards the inside and outside of the building. This can be an onerous responsibility. If he fails to do this the owner of the building usually will have the right to carry out the maintenance himself and then to recover the costs of this from the tenant.

Major maintenance work, perhaps necessitated by the discovery of a building defect, may itself need careful consideration of the appropriate procurement route for having the work designed and carried out.

The building will thus continue on its life until the next development appraisal causes it to remain, to be adapted, demolished and/or rebuilt.

The cycle of the process of land use, property appraisal, development and construction is shown in Figure 1.1.

Summary

Building is a major enterprise and generally is carried out only when:
- no other building exists, or appears to exist, that meets or appears to meet the needs of a client

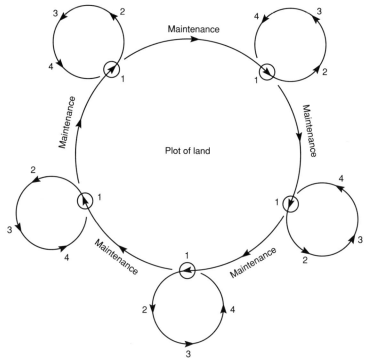

1. Development appraisal
2. Development preparation
3. Development implementation
   design
   demolition
   adaptation
   construction
4. Disposal/retention of property

Figure 1.1   Cycle of land use, development appraisal, demolition, construction, maintenance and reappraisal

- an occupier, an investor or an appointed agent procuring for a client, or sometimes a property development company, will be the client
- a development appraisal shows that the proposed investment will provide an excess of benefits over costs.

2 Deciding to Build

As stated in Chapter 1, new building or renovation/adaptation of an existing building is necessary only when no other building exists, or appears to exist, that either will meet or appears likely to meet the needs of the client.

A fundamental publication (because it gave simplified guidance from the customer's, from the client's point of view) in the area of building procurement was *Thinking about Building* (*TAB*) published originally by The National Economic Development Office (NEDO) in 1985. This publication was a development of evidence and conclusions given in the Building research report *Faster Building for Industry* (*FBFI*) 1983, NEDO.

NEDO no longer exists and the revised edition of this book takes into account modifications made to *Thinking about Building* made by the Construction Round Table (CRT) and now published by The Business Round Table Ltd. The author is indebted to The Business Round Table for permission to refer to *Thinking about Building* and in particular to follow broadly the identification of client's priorities and the classification paths used in *TAB*.

Seven steps to successful building procurement

The CRT/*TAB* guide identified seven steps to successful building procurement:

1. Selecting an in-house project executive
2. Seeking advice
3. Defining requirements to build
4. Timing the project
5. Selecting the procurement path
6. Choosing the construction professionals
7. Choosing a site.

These are discussed in detail below.

Procurement strategy

The first problem is to define the problem. This quotation of HRH the Duke of Edinburgh appeared for some time in the foyer of the offices of the government Property Services Agency (now privatised), Block C, Wellesley Road, Croydon. It is a reminder that the most important decisions are made at the beginning of projects and generally thereafter each subsequent decision is likely to have less and less effect on the overall outcome of the enterprise.

Selecting an in-house project executive

Not surprisingly research has confirmed that organisations that build regularly have people with developed project management skills, that is, people employed inside the organisation. In this context, at the dawn of a probable project, the project management skills required are those of experience of the total process of planning, designing and constructing buildings together with experience of and authority within the client's organisation. The skills required go beyond those of managing a construction project once on site and start before any decision even to build has been taken.

For a project of any size and/or complexity a nominated in-house project executive (sometimes also called a 'project sponsor') should:

- be available full-time
- be the single-point contact for the organisation
- at the very least be able to answer all incoming questions fully and promptly
- understand and organise the internal decision-making processes required for the project
- have the power to speak and act for the organisation
- act in support of any external project leadership appointment.

Organisations that generally have such executives would include the principal government departments or agencies, property development companies, major industrial and commercial organisations that regularly build facilities such as factories, offices and so on. Understandably the quality of executives will vary between organisations, and also within an organisation, but it is fundamental to the success of a project that the in-house executive responsible for overall procurement is equal to his task.

This leads naturally to consideration of a project when an appropriate in-house executive with sufficient skills and/or sufficient time is not available. This returns to the Duke of Edinburgh's quotation and leads to the role of a 'principal adviser' or 'client's representative', a role for which, potentially, the demand is growing.

Seeking advice

Capital investment in a building will not usually be decided without board approval and all the skills of reporting, communicating and obtaining timely decisions to and from 'board level' will be vital for successful procurement.

If the organisation exploring the need to build is an experienced procurer of building it is fundamental to question why a suitable in-house executive is not or cannot be made available. If the project is of any significance to the organisation it must surely be most appropriate for an insider to carry out the role set down earlier. Principally the insider's role is one of understanding sufficient of the organisation's objectives and then organising the internal decision-making processes of the enterprise itself.

If the client does not have an in-house executive with the time and/or skills to run a building project this must be recognised, along with the reasons for it. Even if the client is an experienced, regular customer of the construction industry the process of procurement may still often remain complex but to an inexperienced, novice customer building procurement will probably appear very complicated and even become stressful. A principal adviser should bring knowledge where ignorance exists and comfort where concern may arise, so making the experience for the client as equable as possible.

Many types of organisation involved in the building process could provide a principal adviser: consultancies – architects, surveyors, engineers, project managers – or contracting companies with design and management skills in addition to a construction record.

The quality of the person or organisation appointed as principal adviser will probably be crucial to the success of a project. If clients are inexperienced then the adviser must gain their confidence. In this way an adviser must ensure that his client understands sufficient of his advice, if timely decisions are to be made.

The qualities required of a hired principal adviser include being able to:

- understand clearly, or be able to learn quickly, about a business, its aims and its priorities
- gain the trust of a client in spending his money.

The role of a principal adviser will include being able to act as the 'client's representative' in:

- complementing skills available within the client organisation
- supplying impartial advice on the need (or not as the case may be) to build and how to go about building.

The NEDO report *Faster Building for Commerce*, 1988, recommended (paragraph 5.4) that services available from members of professional and trade bodies should include that of an 'independent customer representative'. But how this service is obtained is not explained.

It is essential that the advice should be impartial, and the client should see it as such. Therefore it is unlikely that the role of principal adviser can be combined successfully with that of any other consultancy or construction service (design, cost, management or construction) that is or will be provided on the project. This is explored further in Chapter 6 – 'Contracts and Conditions'.

Defining requirements to build

As any customer knows, other things being equal, you can probably have what you want if you know what that is – but as often happens if you do not know your exact requirements very probably you will end up with partly or totally what you do not want.

In building, recognised procedures and 'plans of work', for instance the Royal Institute of British Architects (RIBA) *Plan of Work*, the Department of the Environment, Property Services Agency (PSA) *Plan of Work* and the Department of Health *Concode*, have codified the process of design and construction. Experienced clients and practitioners will be familiar with these. Newer bodies, such as the British Property Federation (BPF), have now produced other manuals. The *Manual of the BPF System*, 1983, details a system for the design and construction of buildings and it, incidentally, includes the responsibilities of a 'client's representative', who is the BPF equivalent of a project manager. A principal adviser, in assisting the client to decide his requirements, should have a good working knowledge of recognised procedures such as those described.

Construction students and inexperienced clients of the building industry also need to understand that the early days of construction projects often have an iterative nature to them, that is, they quite often go over ground again and again, in order to establish the client's requirements, including the option, which should always be answered, 'does the client need to build at all?'

'Briefmaking', the radical, searching process that a client and his advisers, particularly the in-house executive and/or the principal adviser, must go through, is needed to explore and to conclude on the nature of the client's business in order to decide why it does or does not require new or renovated building.

The Latham Report recommended that further work be carried out and working groups have been investigating:

- briefing for clients
- a code of practice for clients of the construction industry
- a 30 per cent reduction in real construction costs.

I have commented before that previous efforts of the industry and of its clients seem to have fallen on too many occasions well below par if the Latham Report needs to have recommended such fundamental reviews and reappraisals, yet again, of basic procedures such as briefmaking and client strategy.

Building clients who are property investors or developers generally know this very well and the equation of development value and cost will be constantly present in the briefmaking stage. The asset value of the finished space should always be borne in mind and features which may reduce the ability to fund, to market or to sell the building should be analysed carefully. The value of something should usually exceed the cost of its provision. For this reason cost and investment advisers, agents, designers, maybe construction advisers, may all need to be involved in the stage of 'deciding the client's requirements'.

Inexperienced clients, by their nature, will need advisers that can draw out from them, during 'briefmaking', the nature of the client's business and why, how and when more building is necessary for that business.

Care in briefmaking is sound counsel because later changes to detailed design or, worse still, changes during the construction stage are generally unsound business.

Timing the project

The whole attitude to time and to the programme required for the design and construction of buildings has changed significantly since the late 1970s. In simple terms, this has been the reaction to a general quickening in economic activity, to the effects of international competition on commerce and industry, to the volatility and change of markets that require a facility to 'come on stream' quickly (and perhaps also to become outdated almost as quickly). The realisation that time, not least because of funding costs, is literally money has caused developers and investors to have a greater interest in faster building, that is, faster than the perceived slower building of a decade or so ago.

In Chapters 4 and 5 the priorities and options for procurement are explored in detail. It is emphasised at this stage that the programme of a project, and the time it will take to achieve, is generally of fundamental importance and must be established realistically as near to the outset of briefmaking as possible. Producing a superb facility for a product, with everything in the facility that could possibly be needed, but a year or so after the product can be sold is obviously bad procurement.

Selecting the procurement path

Research carried out for NEDO, culminating in the *Faster Building for Industry* report, led to conclusions on how best to use the construction industry, which 'game plan' or procurement path was applicable and under which circumstances.

The key to procurement is to identify the priorities in the objectives of the client and to plan a path, a procurement route, that will be the most appropriate. It is emphasised that priorities must be put in order of precedence, each in order before others, because by definition there can be only one priority. The in-house executive and/or the principal adviser will be crucial to supply experience of many building projects on which to base a choice of procurement path. The available procurement options are described in detail in Chapter 4 and choosing the procurement route is described in Chapter 5.

Choosing the construction professionals

The only fundamental reason to employ anyone or any organisation to help in procuring a building is because that employment can provide a service or product which is better and/or cheaper than the client himself can provide. Regular users of the construction industry quite often provide some of the skills themselves. Examples of these are government agencies such as the Property Services Agency (now privatised), the Department of Health and local government authorities that directly employ skills of briefmaking, project management, design and cost consultancies. Examples in private industry are property companies and large retailing organisations that often also employ directly a range of similar skills.

The *aspect* to the service or the product being obtained better and/or cheaper than the client himself can provide is worth noting. To measure 'better' or 'cheaper' may not be as easy as it appears and very quickly judgement of 'value' will become necessary. Methods of measurement, terminology, assessment criteria, accountancy, how figures are made up, how overheads are allocated and so on will perhaps need to be debated. It is perhaps for this reason that a number of large property/construction companies have or have not carried out their own construction, why government agencies do or do not provide their own consultancy services, do or do not put them out to private consultancy. Assessment over the short and over the long term may produce different conclusions. For instance, according to how promotion, marketing, depreciation, overheads, goodwill of a practice, taxation, long-term strategy and sinking fund provisions for it are taken into account, may each influence the way in which costs are determined.

If it is then accepted that even the most regular purchasers of construction services and products do not, and would not wish to, carry in-house enough skills always to be self-sufficient it should also be accepted that the manner of purchasing those skills is often most important to the client. The nature of the skill that is lacking is the key to the assistance that must be bought. The lack of the skill may be absolute or it may be partial. There may be no architectural skills or there may be no bricklayers in the client's company. It may be that the client has very good in-house project management skills but cannot afford to deploy them on a particular project.

As shown in Chapters 4 and 5, different procurement paths will involve different responsibilities, risk and workload for a client. Client responsibility may bring with it a degree of monitoring and control that may possibly affect the risk of the client and increase or decrease his workload. Choosing people and organisations to work for the client is critical to successful procurement and it is difficult, in outlining how to choose people, to avoid what may appear to be trite statements. But the following are crucial:

- recognition of the lack of a skill is cardinal – obtaining it is the key
- recognition of vested interests – people and organisations naturally wish to sell their services
- responsibility for any action stays with the person who has it until he specifically delegates it – if it is important, delegation should be in writing
- attitudes and interpersonal relationships are fundamental – match the client's with others' attitudes
- organisations consist of people – make sure the people who will actually work on the project have the appropriate attitudes, the personal qualities, including capability – decide on the people to be involved, not just on the principals with the name
- mutual trust is essential – speed of operation can then flow, quality and value can be achieved
- construction contracts seem to be growing longer in their written conditions, on more and more occasions. Harshness, penalties and litigation have increased and naturally the involvement of lawyers in all this has increased (which came first in the cycle of events is an interesting debate). Cooperation and fairness are vital if a building is ever to be procured to programme, to price and to product quality, and contracts should encourage positive traits, not undermine them with harsh conditions. The use of standard contracts, unamended, remains one way of seeking a balance and a familiarity in the placing of construction work

- price competition is important and is here to stay as the back-cloth for both contractors and consultants – but it is not usually the only or necessarily the main selection criterion
- the most crucial appointment is the principal adviser – obtain him by:
  - comparing several firms
  - seeing the person who will be in day-to-day charge
  - checking references and track record
  - accepting his qualities and that the relationship of trust between client and principal adviser must be right. He must rapidly become an extension of the client's own organisation
  - putting the price of his services into context with the benefits he should bring
  - judging the impartiality and integrity he will have. His advice must not be given simply to maximise, or even to minimise, the workload of his or of any other organisation.

Choosing a site (or building for renovation)

After the previous points this may appear to be out of sequence. It appears here by way of a postscript, a warning to inexperienced clients, students or practitioners.

The NEDO *Faster Building* report showed that quite naturally delays and expense to the overall building procurement time for a client's building often resulted when a client had already determined his site or had already chosen a building to be renovated, before that site or building had itself been professionally appraised. To experienced clients and practitioners this error is obvious but it can happen all too often. Basic studies of statutory requirements, planning, traffic, environmental impact, public utility supplies, soil and structural conditions should be carried out before any commitment to a site of building is made. A principal adviser and a professional team should survey and recommend suitability as part of the client's briefing process referred to above.

Summary

Before deciding to build a client should:

- make an in-house executive responsible for the project
- bring in an outside, principal adviser to give independent advice
- take great care in defining needs and priorities for the project
- select the procurement path which best fits the priorities

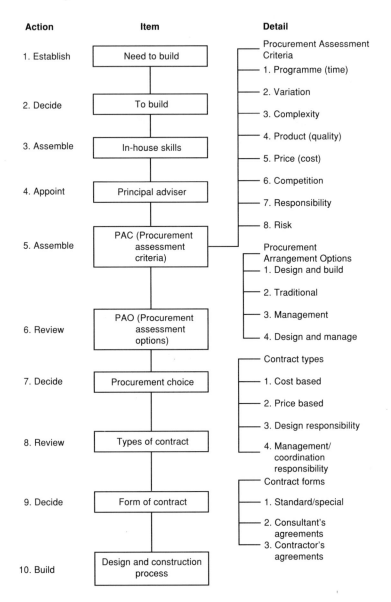

Figure 2.1   Building procurement process

- take considerable trouble to select people and organisations to work for the client
- not commit to using a particular site or renovating a particular building until all options have been explored.

At the moment that a decision is taken to commence the building procurement process a client should know, fairly realistically:

- PRODUCT – HOW GOOD – what the quality of the building will be
- PROGRAMME – HOW SOON – when the building will be available
- PRICE – HOW MUCH – what the building will cost.

The building procurement process is shown in Figure 2.1 and in subsequent chapters the relationship of parties, procurement assessment criteria (PAC), procurement arrangement options (PAO) and types and forms of contracts are explained.

3 Relationships of Parties

Construction services are provided in different ways in countries throughout the world. Different patterns of working have evolved in each country, whereby designers, constructors, suppliers and others relate in different ways to provide buildings that are a constituent of property.

In construction procurement, the relationship of parties involved in the process may vary, according to the procurement route chosen. The client, varied though he may be as shown in Chapter 1, will always remain the client but the various consultants and contractors involved may or may not have a direct contract with him and their duties will probably be different, from one procurement system to another.

How the parties evolved

Before understanding the present-day relationship of parties involved in the construction industry in the UK it is necessary to review the historical origins of UK construction and the way the parties themselves have evolved over time, to reach their present state.

Records of building from the Middle Ages show that the master mason was the contractor of his day. Because buildings of any significance were built in stone, a mason became responsible for engaging and organising men and material to construct a building. It should be remembered that building was a slow process and a major castle, church or cathedral may have taken easily a century or more to construct. Labour was relatively inexpensive and materials were limited to mainly timber and stone. The construction was carried out on the basis of a client's outline idea. There were client's representatives, some with little experience of building, sometimes with a title of surveyor or clerk of works. In an age of illiteracy these men were distinguished by at least some degree of literacy and numeracy. The client, with their agreement, would pay directly for the labour and material used – in effect what is known today as a prime cost reimbursement contract.

The prime cost basis was common until the seventeenth and eighteenth centuries, when a 'bargain' or contract basis began. This was an agreement on a price for work before it was carried out. Master masons had become more important and they gave advice on a number of projects, in a way which resembled that of an architect. Such limited engineering skills as were then required were mainly developed and retained by master craftsmen.

The move towards a contract, 'bargain' basis continued and was given particular impetus by the Great Fire of London. In an attempt to control the cost of the rebuilding that followed the fire, the 'measure and value' basis of payment was adopted. Naturally with this came the employment of separate measurers. There was also dissatisfaction by government at the system of letting and controlling building work (how many official reports and studies have reviewed this subject since then?). The emphasis in contracts and payment for building work was therefore gradually changing from one of agreement and payment for labour and materials on a prime cost, daywork basis, after the work had been done, to one of measurement and valuation of work in advance by agreeing a 'bargain' or contract, before the building work was carried out.

The Industrial Revolution made Britain a trading nation leading to travel throughout much of the world and a great increase in industrial activity and therefore in construction activity. As a result of travel, particularly the European tour, even before the Industrial Revolution, interest was awakened in the property classes and others in the classical buildings of Greece and Rome. The role of the architect, who could study and specialise in design as opposed to construction, became developed here and abroad. Architects sometimes also then acted as developers.

During the very great increase in construction, brought about by the Industrial Revolution, engineers of several types emerged. Civil engineering works, in railways, bridges, viaducts, aqueducts and associated projects, that incidentally did not usually employ an architect, required however specialist engineering and administration skills. As sanitation, heating, ventilation and lighting moved from being viewed as luxury options to becoming very often necessity requirements, the engineering knowledge required to design these facilities needed to expand and there was an increase in consultant design practices, separately skilled in those areas. Interestingly, the expansion in civil engineering construction also brought about a number of engineering contracting companies that carried out both design and construction, in competition with the consultancy-only design practices. Quantity surveying skills became firmly identified and were increasingly provided by separate organisations, having moved from a base of providing bills of quantities to a number of contractors to a position of being recommended by an architect and then, latterly, sometimes being separately appointed by clients.

Specialisation in providing design skills increased in response to the availability and complexity of new building materials (first iron, then steel, reinforced concrete and so on) and also as a response to new technologies (public health sanitation engineering, air treatment in the broadest sense, communication, protection and security systems and so on). Naturally clubs, trade associations and professional societies were established to discuss common problems, and some time later to set qualification standards for membership of those clubs and societies. Architectural clubs came about by the turn of the eighteenth century, civil engineers' organisations even before that. Towards the mid-nineteenth century, surveyors and builders' organisations were formed. These clubs developed, in the Victorian era of the professional man, into professional institutions that defined their area of professional expertise, so creating a public image. Virtually all such institutions sought and obtained a royal charter. Education and training standards, and permission to become a corporate, chartered member of one of the institutions or institutes, was administered by the institutions themselves.

Engineers were involved, generally without architects, in industrial factory buildings but were also employed to advise on the structure of architect-designed buildings. So architects were dependent, technically, on engineers but engineers were not dependent on architects. The separation of architects from engineers in the UK was perhaps a product of our educational system. Architecture was, perhaps still is, seen as one of the arts, whereas engineering and construction were seen as part of science, commerce and industry. Architects were primarily seen to be interested and concerned with buildings of eminence, with prestige projects. As the separation of architects from engineers was evolving, separation from builders was being formalised. Architects were forbidden from practising as an architect at the same time as taking profit from being a director of a building company.

While the institutional, professional part of the construction industry was evolving along separate, specialist, skilled paths the building construction companies were responding to the enormous amount of construction work. Some of today's well-known construction company names can trace their history back to the nineteenth century or before. Specialist engineering companies had also been formed to respond to requirements for piling, for reinforced, often precast, concrete structures, for steel structures and gradually, since the early 1900s, for air conditioning.

The rise of the general building company continued until very large companies existed that carried out virtually all construction themselves, by providing the majority of labour and materials. They provided apprenticeships, employed large labour forces and were 'builders', that is, they employed directly all the building trades and carried out the work themselves, before the transformation to their becoming 'contractors'.

The rise of the 'general contractor' was paralleled, in the years after the Second World War, by the emergence of specialist companies. This

tract, now in a revised edition at the end of 1995 called the NEC Engineering and Construction Contract (ECC), should be used as a basis for this change. Enthusiasm for this has been slow to form and little change seems likely in the industry, apart from use of this form by a number of clients, for instance by British Airports Authority on works for the underground baggage-handling tunnel at Heathrow airport. (This new form of contract is discussed further in Chapter 6)

- contract drafting organisations – The Joint Contracts Tribunal and the Conditions of Contract Standing Joint Committee should coordinate their work and eventually merge. Neither proposal seems to have progressed very far
- disputes – Adjudication should replace arbitration and litigation which could only be available after project completion. The Housing Grants Construction and Regeneration Act 1996 contains provisions for adjudication as a pre-requisite for dispute resolution
- partnering – This should be considered so that it contributes as a means of procurement alongside the existing methods
- better client guidance on briefing should be drafted – This has been done and is expected to become public later in 1996
- a Construction Code of Practice – This is being produced to inform and advise public sector clients (and be available for the use of private sector clients) and should be available later in 1996
- contract standardisation should be developed with a 'family of documents' produced to include virtually all forms of contract between client, contractor and consultant – This very large task is being considered. Attempts to standardise consultants' contracts have met with mixed enthusiasm and institutions have continued to publish their own guidance and conditions of contract, although the Construction Industry Council has published a standard form intended to be capable of use by any professional. The Joint Contracts Tribunal has prepared drafts for a consultancy agreement, common services and procedures
- project managers and project sponsors should be more widely used by clients. Definition of roles and responsibilities for these is the first thing necessary
- good practice in selection of contractors, subcontractors, suppliers and consultants should be encouraged within a framework of 'good practice'. This is accepted but implementing procedures to make it more widespread has yet to be agreed.

I have amended my first edition text for some of the sections that follow in the light of some of the above.

The client

The client and the categories of client have been described in Chapter 1.

In Chapter 2, the importance of the appointment of an in-house executive within the client body for a building project was discussed and it is emphasised again in this chapter. In discussing relationships of the other team members it is vital to understand that the relative success or relative failure of a building project depends on the degree of involvement and the role that the client requires to play and how effectively he does or does not play that role. This has been emphasised again throughout the Latham review.

In the relationships that the client has with consultants, contractors, statutory authorities and other relevant parties, the success that the client (that is, the 'in-house executive') has will largely depend on:

- the structure of his organisation
- the authority vested in him
- his personal characteristics
- his knowledge and experience of the building procurement process
- his successful assembly of a group, a team, of designers and constructors, including most importantly a 'principal adviser'.

Although *Thinking about Building* envisages a position when an in-house executive is perhaps not appointed (that is, when a principal adviser is appointed to carry out some of the in-house functions, plus ones of project management) it is not easy to envisage such a situation for a project of any significance, nor is it easy to speculate that the missing link of a dedicated in-house executive will not be critical to the success of a project.

Clients are now represented at national level, as indicated and discussed previously, by the Construction Clients' Forum. Other client bodies such as the Construction Round Table are committed to making a contribution to the vital task of improving client understanding and performance. The occasional client remains the one needing most help.

The design and construction team

The UK construction industry is now represented by two main groups that combine as 'a team' to produce a building:

- *consultants* that offer design and cost control services and are independent of any commercial interest in construction companies or property companies. The design function may need to include contributions from architects, civil, mechanical, electrical and plumbing engineers, quantity surveyors/financial and cost consultants, landscape architects, interior and graphic designers, process and production engineers. At national level consultants, whilst each category of which are probably members of their respective professional institutions, are represented by the Construction Industry Council (CIC) which in turn is represented on the Construction Industry Board

- *contractors* that are essentially commercial companies that contract to construct buildings. There are many specialist firms of craftsmen and suppliers that manufacture and install the constituent parts of a building, and work as subcontractors to the main contractors. Examples of these are lift manufacturers, air-conditioning companies, specialist roofing companies. At national level companies as described above are represented collectively on the CIB by bodies such as the Construction Industry Employers Council (CIEC), and Constructors Liaison Group (CLG).

A feature of the UK construction procurement process is that generally each project allows a diversity of choice in setting up the consultancy and constructor relationships on that project. Because of the choices available, particularly among the construction elements, and because of the presence of competition, design and construction teams in their totality probably do not repeat their roles very often. Relationships are quite often formed for the first and only time on a project.

Because of this Latham and others have proposed that other methods of forming relationships should be used from time to time. One such method is 'partnering'. This is already used more extensively in other parts of industry and commerce. Partnering means different things to different people. I am taking it to mean any arrangement whereby people are encouraged to work more efficiently together. Partnering has a number of ingredients, according to the Centre for Strategic Studies in Construction at Reading University, including shared problem resolution, continuous improvement, reliable product quality, continuity of work, fast construction, completion on time, lower legal costs and improved profits. This may be achieved by forming longer-term relationships between clients and their consultants and contractors or between contractors, their subcontractors and suppliers or it may be achieved by fostering these objectives on a 'project by project' basis. For instance, instead of appointing a consultant or contractor, following perhaps the usual procedures of competitive selective tendering, a client will negotiate with a company which is one of a small number of companies with whom the client regularly, and perhaps exclusively, works. A post-Latham work group is researching means of achieving 'partnering' and is due to report later this year. Partnering's supporters are enthusiastic about the benefits but admit that they can only be achieved through good management skills, shared objectives and continuous will to resolve potential conflicts almost before they arise.

Clients such as British Airports Authority have instituted extensive reviews of project managers, architects, surveyors and so on to establish a limited number of organisations with whom, say over the next five years, they will contract, including by 'partnering.' 'Partnering' is seen to be of benefit to some clients and also to their 'procurement suppliers' by ensuring an efficient, and probably regular, source of work between a limited number of organisations who should therefore work more effectively together. This should result in the provision of better quality and value for money for both parties.

The traditional structure of team relationships, leaving aside project management (which is referred to later), being placed between the design team and the client, is largely (other than for design and build procurement) one where the architect is responsible for helping to prepare a brief, then continuing for design and management of the project through the design stages by coordinating the work of other design consultancies, through the construction period until project completion and the handover of a completed building to the client.

Each design and cost consultancy contribution will generally come from separate professional practices with whom the client is in direct contract (Chapter 6 discusses these contracts, generally termed 'conditions of engagement', and the services that they offer). The interdependence of each consultancy, one with another, working on the one project, is obvious. The more complex the project, the more interdependent are the consultants likely to be. Any unsatisfactory contribution by one member in a design team is most likely to affect the contribution of other members.

The architect often has to perform two roles, first that of designing a building and second that of administering the project. He does this by coordinating the contribution of consultants and then administering a building contract – not managing the construction process. Managing construction is done by a conventional contractor, or by a management contractor or construction manager, or by a design and build contractor (there are variations on these, as described in Chapter 4). In performing two roles, the architect may, on occasion, not easily be able to exercise objectivity in his decisions. Not many persons easily combine the skills of design of a project with those of management of that project, or with the requirement to carry out quasi-arbitral roles under a building contract, especially when the cause of some of the claims of a contractor may be due to the underperformance of the architect himself.

Because a dual role is expected of architects and engineers under many building contracts, provisions have now been made in recent forms of contract for separation of the role of designer from that of the contract administrator. This is referred to in more detail under 'project manager/management' below.

Integrated design teams

Integration within a design team is often difficult to achieve in the traditional structure because the objectives of each organisation, which ideally should be the same as those of the client, may not always coincide. Each consultant generally has a separate contract with the client, the terms of which are sometimes incompatible with some of the duties contained in the contracts of the other consultants. Because, in the traditional structure, the integration of the team has not taken place often enough, initiatives have been made to appoint a 'lead consultant' who would himself take responsibility for 'leading the team' and may even be asked to appoint the other members of the design team as subconsultants to his consultancy.

In order to reduce divisions of responsibility some clients have begun to appoint on a 'lead consultancy' basis and some consultancies now offer, either through consortia of separate organisations or through integrated, multidiscipline, single-practice consultancies, the provision of all design (and sometimes cost) consultancies.

The Ministry of Defence (MoD), among others, makes many building consultancy appointments on the basis of a 'lead consultancy' commission. The organisation tendering to offer a range of services must enter into subcommissions with others if it cannot provide all the services from its own resources. The MoD also requires to know the 'subconsultants' proposed to do the work.

It is interesting to note that the quantity surveyor has generally promoted his 'independence' and increasingly has sought to become appointed before other consultants. Even if multidiscipline services are provided under one contract agreement, it currently seems unusual that quantity surveying/cost consultancy services would be part of that agreement. It seems that, as with the appointment of a project manager before the appointment of a design/construction team, the appointment of a cost consultant before the appointment of the design consultants is seen often, as with project management, as a 'watchdog', 'super consultant – reporting-outside-the-team' role.

The whole discussion of how much design, cost and construction consultancies can and should be combined within one organisation, and are therefore readily able to offer one contract with the client, remains unresolved. An organisation should theoretically have the fewest divisions as a team and be able to provide the greatest opportunities and evidence of full integration if it combines all the functions of design, estimation, cost control and construction and in addition be part of the client body. Whilst in theory this is so, conflicts in the objectives of individuals can make the practice more difficult than the theory. Contractors/developers quite often do not carry through development projects using their construction arm and local authorities generally find justification for direct works procurement very, very difficult. Togetherness of itself is usually not enough and the separate development of professionals and constructors described earlier needs to be managed and formed into one organisation with a common objective.

Project manager/management

Probably because of the evolution of separate design and cost consultancies, contractors, subcontractors, specialists and so on, the project manager arrived, in the mid 1970s, and perhaps has been accepted by some. It is argued that this is because the architect and other key members of the design team have failed to provide essential 'management' in order to coordinate the overall process of planning, design and construction and have therefore allowed a new profession, that of project management, to become necessary. Project management as a skill and as a function is reasonably easy to define and to want but how best to achieve it is still debated quite strongly.

It is necessary to define 'project management' and 'project manager' because they are terms that are used in several ways. In this chapter project manager does not mean a contractor's site or contract manager (contractors quite properly use the term 'project manager' to describe the person who manages their contract, their project). Likewise it does not mean the person used by estate agents, property companies and property developers to manage their developments.

The terms 'project management' is defined in *Project Management in Building*, published by the Chartered Institute of Building, 1988, as meaning 'the overall planning, control and co-ordination of a project from inception to completion aimed at meeting a client's requirements and ensuring completion on time, within cost and to required quality standards'. It is interesting that the term 'client's representative' has also emerged for use in circumstances where a client appoints 'his man' to oversee a function that cannot, it is argued, be performed by any of the other parties.

The official reports of Banwell and Simon, referred to above, virtually all stressed that more attention should be given to the way projects were structured and how the structures were to be managed. 'Management' was seen as a skill in its own right that was not being provided, as it were, by accident. Management was perhaps the missing ingredient that, if injected, would put right many of the deficiencies of the construction industry found in the official studies. Project management was a function recognised in other industries and in industry generally, for instance, in the oil and heavy engineering industry where project management systems and project managers are regularly used.

Latham has made recommendations for better management of projects by reference to 'project sponsors' and to 'project managers'. He said that 'before deciding on their contract strategy clients should assess if they need a separate project manager as well as an internal project sponsor or whether another procurement route should be followed such as design and construct or the use of a lead manager who also acts in a design and/or a supervisory role'. This would seem to be covering virtually all possibilities. But he did go on to say 'if a separate project manager is necessary'. Because of the dual role expected of architects and engineers under most building contracts, provisions have been made in other forms of contract for separation of the role of designer from that of contract administrator. In particular the NEC Engineering and Construction Contract (ECC), referred to in more detail in Chapter 6, has separated the design and management roles. In the construction contract there is a project manager, a supervisor and of course a client (the 'employer') and a contractor. It is envisaged that a client will have 'designers' but that his project manager will 'have to manage their activities' (a quote from the guidance document issued with the ECC). Public sector contracts, in the suite of forms issued under GC/Works/1 conditions, also refer to a project manger. It is a growing trend, certainly for larger building projects, to have a project manager and for him to be the administrator of the project, managing both the design and the construction teams.

It is interesting to see not only how project management is developing but also that project coordination is being provided. Words as often may say it all. Criticism of the construction process was, and still is, of the lack of liaison between architects and other professions, between designers and contractors, between all of these and their clients. The responsibility of a project manager to 'manage', to guarantee and improve all of this is an onerous responsibility. The responsibility of someone to 'coordinate' this may be a good deal less.

The construction industry continues to evolve aided by self-examination and prodded by reports and studies. The evolution has been given an impetus by the realisation that a project manager may become more than a manager – he may become very influential in the way that the 'principal adviser', referred to by CRT in *Thinking about Building*, is envisaged. The position has now been reached where many organisations purport to offer project management or certainly project coordination (perhaps because it may carry less liability). Architects, engineers, quantity surveyors, construction advisers and project managers have each pursued their own approach to project management, while stating all the time that the role of project manager is not the given right of any one profession or construction adviser (whilst of course, by coincidence, often stating that their own background is the most appropriate one from which to provide objective project management). Management (or the lack of it) has been seen throughout large parts of UK industry and society as being the area which, if improved, would lead to marvellous benefits. In the construction industry, the emergence of management contracting, construction management, systems management, alternative methods of management and so on have been part of the national response to improving management, to bring it up to the best of international standards.

The techniques of project control such as critical path analysis, programme and cash flow forecasting, labour and material resource allocation, cost in use appraisal and so on have become available, but mainly in an uncoordinated way. These are techniques and have been taken up by many professionals and construction advisers but their application has been piecemeal and inconsistent across the industry, perhaps because some clients find some techniques of use and other clients do not. No consistent implementation of project management and project management techniques on a universal basis seems to have occurred, perhaps because there have been inconsistent requirements from clients, no universally accepted framework of project management, nor a very generous or hospitable acceptance by the professions in general that 'project managers' are necessary or that they add value. It may well be in reality that they do but that the thinking, training and attitudinal changes required of professionals and constructors also need to move considerably before project management can show a track record of success.

Latham recommended that if the practice of employing project managers is to increase then their roles, duties and terms of appointment should be clearly defined. Evidence of an applicant's necessary practical experience of the construction industry and of specific management skills to carry out his duties should be available. Once appointed, a project manager should be given authority to carry through a project to completion. His terms of appointment should be interlocked with those of the design and construction teams.

An increase in diplomas in project management and attempts to form a generalised, accepted body of knowledge that can identify significant features in the planning, design and construction process suggest that the subject continues to be of growing interest.

Many design professionals now accept that their education and training does not equip them for the management role. Largely for this reason clients such as the PSA (now privatised), the Department of Health and members of the British Property Federation have tried innovations in managing design and construction teams and often introduce their own project manager or 'project sponsor'. This is recognising the necessary 'in-house executive' role (referred to in *Thinking about Building*) but it is not recognising the need for independent project management organisations and Latham has addressed this. The experience of project management across a whole range of projects is insufficient from which to generalise, except that larger projects increasingly seem to involve project managers and that the advantages claimed in the control of time and cost have not always been adequately researched and/or proven.

A prime difficulty in judging the success, failure and therefore worth of project management is in establishing exactly the responsibilities of a project manager but this should come out of the post-Latham work in this area. Definition of responsibilities for items such as completion on time, cost and budgetary control and quality standards are not easy and responsibility depends on authority, otherwise the responsibility is hedged by terms such as 'advising on', 'monitoring', 'evaluating', 'anticipating' and 'coordinating'. Many observers continue to say that a project manager, in these days of litigation, dare not himself take on many real responsibilities unless he designs, estimates and constructs himself by directly entering into contract with the organisations that perform those duties – this would seem to be 'design and manage', BOOT or whatever. BOOT is an acronym for 'Build Own Operate Transfer' and is described in Chapter 4, under variations on the procurement option of 'design and build'. On the one hand, it is argued that a skilled project manager, by delegation, can obtain the knowledge and advice from the design, cost consultancy and construction team to enable him to manage. On the other hand, many suggest a project manager can, at best, coordinate this advice and manage *some* of its consequences but that he cannot be responsible for it.

The future may see an improvement in the quality of service provided by project managers, generally, and the training and qualification of people specifically to become project managers may continue to raise standards such that the above comments are no longer relevant or at least become exceptional.

Relationship of parties

The contractual relationships of the parties are shown below, according to the procurement method and therefore to the contract type adopted. A special publication No. 113 by The Construction Industry Research and Information Association (CIRIA), titled *Planning to Build*, explores the relationships of parties in different procurement methods.

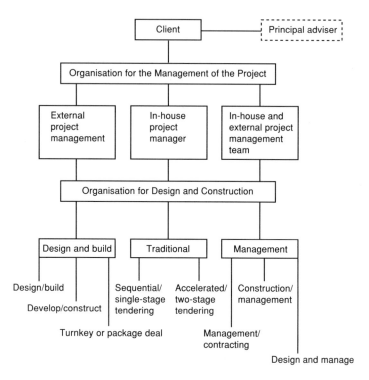

Figure 3.1   Range of organisational variations for project management, design and construction

The overall range of organisational variations for a client is shown in Figure 3.1. The client may or may not have in-house project executive skills, as referred to in Chapter 2, and he may therefore provide these accordingly by one of the three options shown at 'organisation for project management' level. The procurement options of design and build, traditional and management then subdivide as shown. Chapter 6 discusses contracts for engaging professional services, for construction works and for collateral warranties.

Design combined with construction: design and build

In this method of procurement, as shown in Figure 3.2, the client is in contract with:

- a design and build contractor to design (or develop a design) and construct a building
- a consultant(s) to advise on the preparation of the client's requirements, select tenderers and evaluate their submissions and possibly

act as the client's agent (the 'employer's agent' in a design and build contract).

It is accepted that a client may have in-house skills that make the appointment of a consultant(s) unnecessary.

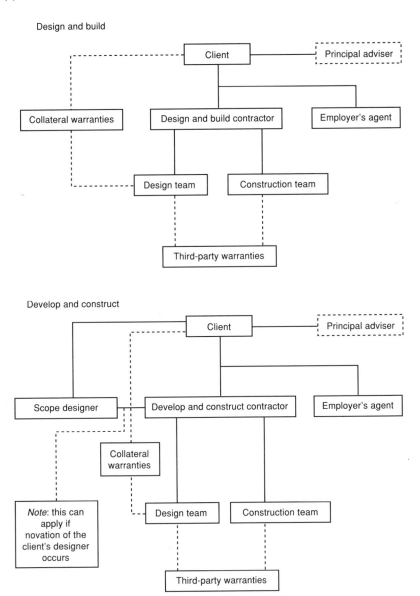

Figure 3.2   Design and build contracting: contractual relationships

The design and build contractor will or may, in addition to his contract with the client, be in contract with:

- domestic suppliers and subcontractors
- consultants for design and/or cost consultancy services (who may also have a contract with the client).

It is less likely but still possible that collateral warranties will be required by the client but they may be required by third parties.

Design combined with construction: design and manage

Contractor
In this method of procurement, as shown in Figure 3.3, the client is in contract with:

- a design and manage contractor
- (possibly) a scope designer.

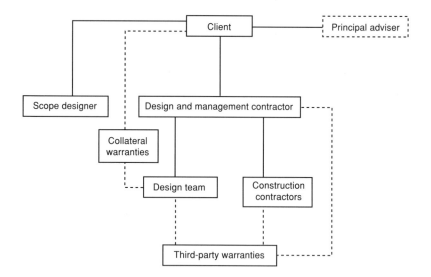

Figure 3.3 Design and manage (contractor): contractual relationships

The design and manage contractor will, in addition to his contract with the client, be in contract with:

- consultants for design and/or cost consultancy services
- works contractors (may be sixty to a hundred organisations).

Collateral warranties may be involved, as described under 'traditional' and 'management contracting'.

Consultant
In this method of procurement, as shown in Figure 3.4, the client is in contract with:

- a design and manage consultant
- (possibly) a scope designer
- a design and/or cost consultancy organisation
- works contractor (may be sixty to a hundred organisations).

Collateral warranties may be involved, as described under 'traditional' and 'management contracting'.

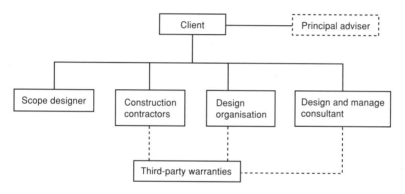

Figure 3.4 Design and manage (consultant): contractual relationships

Design separate from construction: traditional (or 'lump sum')

In this method of procurement, as shown in Figure 3.5, the client is in contract with:

- a contractor to construct the building
- consultants for design and/or cost consultancy services.
- subcontractors through collateral warranties
- suppliers through collateral warranties.

The traditional contractor will or may, in addition to his contract with the client, be in contract with:

- domestic suppliers and subcontractors

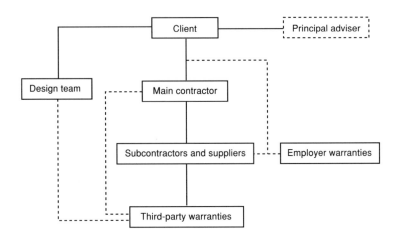

Figure 3.5   Traditional contracting: contractual relationships

- nominated or named suppliers and subcontractors
- a third party through collateral warranty requirements for funding, purchase and tenancies.

Consultants may, in addition to their contract with the client, be in contract through collateral warranties with third parties for funding, purchase and tenancies of the building.

Design separate from construction: fee management

Management contracting
In this method of procurement, as shown in Figure 3.6, the client is in contract with:

- a management contractor
- consultants for design and/or cost consultancy services
- subcontractors through collateral warranties
- suppliers through collateral warranties.

The management contractor will or may, in addition to his contract with the client, be in contract with:

- works contractors (may be sixty to a hundred organisations)
- a third party through collateral requirements for funding, purchase and tenancies.

Management contracting

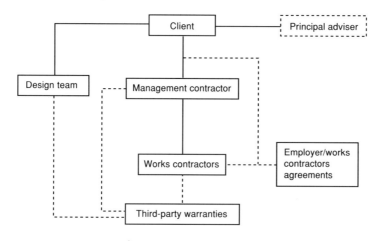

Fee management contracting

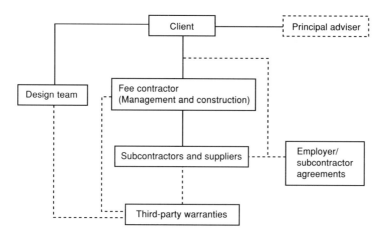

Figure 3.6 Management contracting: contractual relationships

Consultants may, in addition to their contract with the client, be in contract through collateral warranties with third parties for funding, purchase and tenancies of the building.

Construction management
In this method of procurement, as shown in Figure 3.7, the client is in contract with:

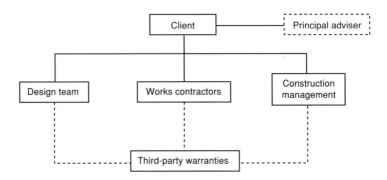

Figure 3.7   Construction management: contractual relationships

- a construction manager
- works contractors and suppliers (may be sixty to a hundred organisations)
- consultants for design and/or cost consultancy services.

The construction manager may, in addition to his contract with the
client, be in contract with a third party through collateral requirements
for funding, purchase and tenancies.

Consultants may, in addition to their contract with the client, be in
contract through collateral warranties with third parties for funding,
purchase and tenancies of the building.

Summary

In considering the relationships of parties the following should be noted:

- historically, each type of consultant and contractor has evolved
  with separate education, training and often attitudes
- the client is central to relationships and will be involved in a variety of contracts, depending on the procurement and contract circumstances
- the Latham Report has confirmed the central role of clients but
  also that many of them are unaware of the traditional relationships
  within the construction industry and often are unable to use the
  industry to their best advantage
- understanding and using the benefits of each type of relationship
  can only be obtained by experience
- in principle, the more relationships that a client has on a project
  the more involved he is likely to become during its progress.

4 Procurement Options

The UK construction industry is flexible in the procurement methods that it can offer and, quite often, the simple categorisation of the procurement options described in this chapter may not be followed exactly. Whilst it is currently accepted that 'design', 'management' and 'construction' can be viewed as discrete parts of the procurement process and, for that reason, categories of procurement and construction contracts have been classified around these three distinctions, it may be that in a later period these 'discretions' may not seem so separate. Combinations of design and construction, design and management, and other variations in the simple, distinct relationship of client, consultants and contractors do now occur as each project's circumstances develop or dictate.

Whilst the client is always 'the employer', which of the remaining constituent parts he employs, that is, with whom he signs a contract or contracts and for which purpose that contract is made, is the variable in the procurement options. The variables of who designs, when the design is carried out, who contracts with whom to construct a building, and so on, produce the optional ways of procurement.

Building contracts and conditions of engagement for consultants that are appropriate to each procurement option are described in Chapter 6. 'Design and build', 'management' and 'traditional' contract forms have been developed to match procurement options *after* each option has evolved and then, sometimes a considerable time after that, the 'standard' forms of the construction industry have each become established. This has been the sequence rather than that building contracts have been produced in advance of the evolution and survival of each procurement option. As an example of this, in mid-1996 no 'standard' form of contract has yet been published for construction management, although this form of 'management' has all but replaced 'management contracting'. It is understood that the Joint Contracts Tribunal has draft proposals for a form for construction management.

Thinking about Building sets down four arguably different ways in which members of the construction industry are brought together to provide a building. In the 1985 edition of *TAB* the four ways for building procurement were illustrated by the use of a picture of a 'ludo board'. In the 1995 edition of *TAB* ideas had moved on, the 'board' illustration had disappeared and instead two main categories for procurement were given – first for 'design combined with construction' and second for 'design separate from construction'. This grouping into two categories reflects a change in attitude over the ten-year span between the two versions of *TAB* and focuses the rise of 'contractor procurement' relative to 'consultant procurement'.

To suit the simplicity of presentation required in *TAB*, a publication particularly for those inexperienced in building procurement, four different ways are perhaps appropriate. For exploration in this chapter the four categories are split further into subcategories. The considerations/characteristics of each procurement option are given in this chapter and it should readily be seen that each option has favourable and unfavourable points – the relative importance of these will lead to the optimum option on any particular scheme. The choice of which procurement option is appropriate to the circumstances of the project is explored in Chapter 5. The balance of priorities will come down to the allocation of the 'risks' in one option over another. A procurement route with most benefits to the client will be the one appropriate to him in the particular circumstances of his project.

It is tempting, perhaps customary, in starting a detailed exploration of procurement options, to use as a datum or a point of departure the 'traditional' system. This would be because 'traditional' is still the most widely used procurement system and probably for that reason it is the best understood, at least by consultants and contractors. In 1996 it still remains the majority choice. The reason why it is the most used procurement route should be only because in the majority of projects it is the appropriate choice, not because it is 'traditional'.

In describing the four main categories of procurement, 'design and build' is however taken first in this chapter. Historically, as discussed in Chapter 3, there are good reasons why the separation of consultancy from contractor has existed fairly rigidly but the separation of design from construction, as functions irrespective of who carries out those functions, is not particularly logical.

In the market sectors of building, such as speculative housing, industrial warehousing and specialist buildings for an industrial process, design and construction have generally been provided by one organisation and finished buildings are therefore often available, ready-made to occupy. It is in most other building sectors that design and construction have been separated, perhaps not always necessarily.

Sir Harold Emmerson stated in the report that he prepared for the Minister of Works (as the title then was), *Survey of Problems Before the Construction Industries*, 1962, HMSO, that 'in no other important industry is the responsibility for design so far removed from the responsibility for production'. The building industry appears very gradually to have taken note of this and the 'design and build' procurement option is now increasingly being used for a significant proportion of projects.

The four procurement options, part of the two groups of options now shown in *Thinking about Building,* are discussed in this chapter in the following order:

- Design combined with construction
  1. Design and build
       Direct
       Competitive
       Develop and construct
       Package deal
       Turnkey
       Private Finance Initiative (PFI)
       Build-Own-Operate-Transfer (BOOT)
  2. Design and manage
       Contractor
       Consultant
- Design separate from construction
  3. Traditional (sometimes called 'lump sum' contracting)
       Sequential design
       Accelerated
       Partial design of parts or elements of the works
  4. Fee construction – the management method
       Management contracting
       Construction management
       Fee management (prime cost contracting)

Design combined with construction: design and build

'Design and build' systems of procurement are not new, although they may appear new to many people today. Architect/builders (for instance Wren) were supplying buildings for their clients (often 'patrons' in those bygone days) before architecture and design became separated from the building process. If the term 'design and build' itself was little used or known until the late 1970s/early 1980s, terms such as *package deal* and *turnkey* were known (if not too well defined). These terms are described later, as variations of design and build, together with PFI and BOOT.

Design and build is the procurement position where one organisation is responsible to the client for both design and construction. Organisations currently supplying the procurement option of 'buying' a finished building are most generally building contractors. Incidentally there is now no logical reason, and virtually no practical reasons, why consultancies could not provide the finished product – but when would a consultancy then become a contractor? Rules for architects, engineers and surveyors practising as companies, perhaps carrying out development and construction work in addition to design are no longer so restrictive. In North America architect/developers and architect/contractors have long existed and this has occurred in a limited number of cases in the UK also.

With the advent of the Private Finance Initiative (PFI), which is design and build taken into a complete package of finance/design/build/lease and/or own, it appears that financial institutions, such as merchant banks, general management consultants and consortia of contractors/consultants, are offering solutions to government agencies or to 'quangos' for initiatives to lease or buy facilities. This is discussed in more detail below.

'Develop and construct', also described and commented on in more detail below, is now used for the majority of design and build procurement. Although design and build has grown considerably in popularity since the mid-1980s it is now claimed by some, mainly contractors, that the system has become too distorted. It has become nearly all 'build' and virtually no 'design'. Jibes of 'R and B', meaning 'risk and build', have become commonplace. More detailed aspects of develop and construct are discussed below.

Components

The components of the design and build system shown in Figure 4.1, are:

- establishing the need to build
- establishing the client's requirements
- selecting and inviting tenderers to bid
- the contractor or contractors preparing their proposals for design, time and cost
- evaluation and acceptance of a tender which then becomes a contract
- design and construction of the works.

The client will need to have in-house skills or to obtain them (see Chapter 2) in order to:

- prepare his 'client's or employer's requirements'
- carry out his responsibilities in the contract or to devolve these to a 'client's or employer's agent'.

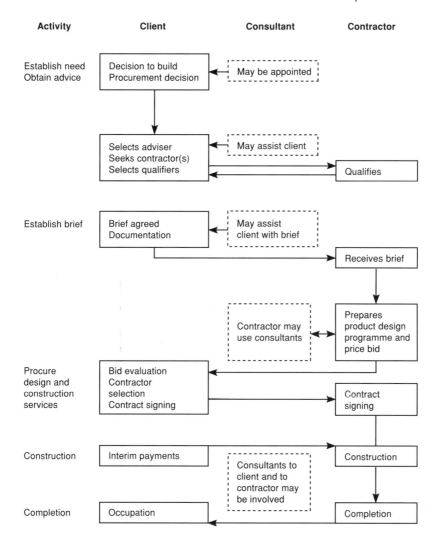

Figure 4.1 Components of the design and build system

Variations on design and build

The common variations of design and build are:

- *Direct*
  In this case no competition is obtained in tenders. Some appraisal of possible competitors may be made before tendering but only one tender is obtained.

- *Competitive*

  Competitive design and build is the most usual procedure with tenders being obtained from documents defining the project prepared to enable several contractors to offer competition in designs and in prices. These documents may be prepared by a consultant or contractor for a fee.

- *Develop and construct*

  Consultants design the building required to a partial stage, often called a 'scope design', then obtain competitive tenders from contractors who develop and complete the design and then construct the building. The amount of consultant design can vary from little more than stipulations over the area of space required, going on perhaps to giving broad indications of external appearance or materials required or, at the other end of the scale, with the consultant preparing a scheme design that specifies all the major components, materials, elevational and fenestration outline, thus leaving the contractor little more than to prepare the working details to enable the building to be erected.

  Tensions in the operation of develop and construct procurement have appeared in the mid-1990s as more procurement has been carried out by this route. Inherent in develop and construct is the flexibility available to a client over how much design is done by his team and therefore how much design remains to be carried out by the develop and construct contractor. Assessment of risk and therefore disputes over the interpretation of this flexibility, both at tender stage and subsequently during construction, can be more considerable than at first may be apparent.

  One of the apparent advantages of develop and construct is that a contractor's expertise in buildability and procurement skills can be used to his and to his client's advantage, potentially bringing economies to both. However some clients do not wish to use develop and construct in this way. A quote from one client has been: 'we employed the contractor for its ability in getting our project built, not for its ability to offer design changes'. This client specified his requirements in great detail and left his contractor with very little scope for 'development'. Ideas and changes suggested by the develop and construct contractor, for whatever reason, were generally not acceptable to the client. Apart from obtaining the important 'one-line responsibility' it is questionable what advantages this client obtained by implementing develop and construct in the way that he did.

  So in many examples in the present industry usage of develop and construct a design team is employed by a client to produce quite detailed drawings from which competitive tenders are then obtained. Then the client novates (that is, asks the contractor to take on the employment of the design team and with it responsibility for all its previous design) that design team to the contractor. This means that a contractor often has had little input into the design and will have little subsequent control or power over that design, whilst being asked to take on all responsibility for it. The present practice has developed because critics of design and build, or develop and construct that is near to design and build, argue that leaving design or significant parts of it to a contractor often means that a client does not obtain the building he wanted or expected. By having a design taken to quite an advanced stage it is hoped that a client's brief is therefore not misinterpreted and that the specification is not lowered after it is passed to a contractor. Novation is generally unwanted by many contractors and it can create problems as well as appear to solve some. Conflicts of interest can occur and 'half-novation', where a client retains the services of his designer to 'check' on the contractor whilst at the same time the designer is employed by the contractor, is considered to be even more unsatisfactory.

Develop and construct with novation is, according to the Reading Design and Build Forum, now accountable for around 80 per cent of design and build contracts, which themselves are probably around 30 to 35 per cent of the construction procurement market (excluding industrial, housing, process and civil engineering).

If a client retains a requirement to 'check and approve' a develop and construct contractor's 'development' the verification process can sometimes involve considerable time and expense, maybe resulting in slippage of construction time, leading to disputes over responsibilities for that slippage and for any consequential costs. On a well-known project for a government agency, multi-million pound claims were submitted partly because, it was said, considerable delay occurred over 'approval' and 'approval time'.

In an age when fierce competition in consultants' design fees has led to accusations of inadequate design services being provided it is hard to obtain facts to form detailed, and therefore definitive, judgements on develop and construct. Design and build has always been criticised for its association with lower quality of product, both in functional and specification levels. This has mainly been said when 'too much' has been thought to have been left to a contractor to design or develop. Disputes have occurred on some well-known, high-profile develop and construct projects over delays said to have occurred during periods of 'approval' of 'development of design'. It is possible that the design and build and develop and construct method of procurement has been pushed further than many of its proponents thought sensible.

There is little doubt that the develop and construct method can place both design teams (that within the control of the client and that within the develop and construct contractor's control) on to a path with a very rapid learning curve. There can be a tendency for the client to say that because design has been developed to a certain stage that the remaining part of design can then quite easily be developed 'in five minutes'. In particular, disputes and delays can often occur during the development stage of engineering services and their installation because of the interactive nature of those services in engineering plant areas, duct sizing and the need for general coordination of mechanical and electrical engineering designs with the structure and the building fabric. In practice develop and construct has not always solved these issues any more successfully than other procurement methods.

- *Package deal*
  The package deal variation is often used where the contractors competing will use a significant part of their own or another's proprietary building system or they will be constructing variations of a repetitive theme. The significance is that *buildings* are provided rather than innovative *designs*. A package dealer may also offer to provide or to find a site, to sell, mortgage or lease his product, to obtain statutory permissions, approvals and so on, perhaps at risk to himself or at a charge to the client.

- *Turnkey*
  'Turnkey' is a relatively old term, originating in the USA, used for earlier examples of 'package deals'. The *key* referred to is perhaps to symbolise the client's only apparent required action, in addition to paying money by means of a lease or outright purchase, that is, to 'turn a key' and take up occupation of his 'buy'. Speculative, private housing is a prime example of 'turnkey'. It is perhaps the simplicity of the word that disguises the exact position because often there is more, sometimes much more, that a client has to do than turn a key and pay money. A client must still go, however briefly, through a briefing process with a 'package dealer'. A client must still make vital decisions in examining and answering the 'why' and the 'how' questions (discussed previously in Chapter 1) – this vital process, however quickly or slowly carried out, is still required.

  However, other matters of finance, design, construction and perhaps finding a site, are put together by the 'package dealer' and thus removed from the concern of a client. More complex products of the building industry are offered by 'turnkey'. Prefabricated buildings for temporary or permanent use can be bought or leased for housing, offices, factories, educational or health purposes and similar examples of fairly simple facilities. More complex facilities can be provided for industrial processes. Arguably it is not easy to categorise 'turnkey' as 'design and build' or (perhaps more correctly) as 'design and manage'.

- *Private Finance Initiative*
  The Private Finance Initiative (PFI) is arguably an example of 'turnkey'.
  PFI was introduced by the Chancellor of the Exchequer, the Rt Hon. Kenneth Clarke, at the CBI Conference in Birmingham, November 1994. In essence he announced that 'the Treasury would not approve any capital projects unless private finance options had been explored'. His aim was 'to maximise the scope for, and use of, private finance while concentrating inevitably finite public capital provision on where, for whatever reason, private finance is not possible'.
  It is fair to say that PFI has not appeared to increase private funds such that more construction projects have been processed, let alone built, following PFI. Several high-profile figures from outside the construction industry have led PFI since 1994, but 'procedures' have appeared to slow projects rather than release more for construction. Major hospitals, prisons, roads, rail links, educational projects and so on are examples of PFI which can be seen as 'turnkey' on a grand scale.

- *Build-Own-Operate-Transfer (BOOT)*
  BOOT projects are similar in concept to PFI and examples could be found before the 'sifting' of public projects came about after PFI. The Channel Tunnel, several major river bridges (the second Severn bridge, Isle of Skye bridge) have already been constructed by financial/construction consortia. The BOOT organisation took a brief from, in the above examples, Eurotunnel or a government agency, and then financed, designed, built, now own (generally for a fixed period of time) and in the meantime (perhaps for fifty years) operate the facility. The 'operate' part is achieved by giving the BOOT organisation a licence to operate the facility for a given period in exchange for having a licence to make charges by way of a toll, or a rent or a 'fee' for the use of the facility that has been provided. The 'transfer' part will come about when the BOOT organisation's term of operation comes to an end and ownership of the facility is then transferred to the commissioning organisation.

Other examples are of 'private' prisons, hospitals and so on which are or will be leased to a government agency which in exchange pays to the BOOT organisation on a basis of 'hospital beds occupied' or 'prison cells occupied' and so on.

Variations on BOOT seem almost endless. An organisation need not be commissioned to own the facility but only to provide the interim financing whilst it is being planned, designed and constructed. As such the 'transfer' will come much sooner than in the examples I have given above. It may or may not be asked to operate the facility it has built (it may, say, only be asked to maintain it whilst others operate it). Advocates of BOOT say that the greatest benefits come from the 'all through' possibilities of BOOT because an operator that has to operate and maintain for, say, fifty years will be very concerned to design and build with operational economy and 'costs in use' to the forefront. Others say that assumptions on finance capital and interest rates (assumed and actual) are by far the most significant factors and that design and construction considerations are marginal in their effect over, say, fifty years.

In describing these developments in building procurement I have chosen to categorise PFI and BOOT under 'Design combined with construction: design and build'. Arguably I could have chosen 'design and manage'. It is apparent that hard and fast lines over whether or not the procurement route is one of PFI or BOOT, or even whether or not the route is 'design and build' or 'design and manage', are increasingly hard to draw. Likewise the relative positions of contractors, consultants and clients in PFI/BOOT arrangements, and indeed whether or not there is any longer any clear difference between the meaning of names given to organisations according to the role they once played in other, apparently more traditional procurement routes, has often become quite difficult to determine.

Considerations/characteristics

- the system can no longer be regarded as relatively new with the 'develop and construct' variation now being very popular
- client obtains single-point responsibility from one organisation – a contractor
- early in the process a financial commitment should be known by the client
- because the contractor has responsibility for both design and construction, it is claimed that this produces economies for both contractor and client

- the contribution of the contractor's knowledge of 'buildability' can be beneficial in programme and price to both client and contractor
- contractors that are familiar with the briefing and design process of specialist types of buildings, for instance chemical and industrial processes, may be able to offer overall programme advantages over alternative procurement routes
- the nature of design and build contracts tends to restrict changes during construction because of the disruptive and relatively high price to the contractor and client
- competition between contractors on both design and/or price may be advantageous to a client – design becomes a competitive element in a tender
- alternative design proposals may not be easily comparable and evaluation may need to 'assess' the incomparable
- design costs of unsuccessful tenderers may be very significant and usually become a cost that needs to be recovered from successful projects
- single or two-stage tendering may be appropriate, thus attempting to limit expensive design costs until a second stage of tendering involving perhaps only one, but certainly fewer tenderers
- the tendering costs of unsuccessful tenderers can be considerable and the NJCC code for design and build recommends that unsuccessful tenderers' costs are reimbursed
- professional advice in preparing employer's requirements and in assessing contractor's proposals may need to be significant in order to select the right tender.

Design combined with construction: design and manage

The 'design and manage' system combines some of the characteristics of 'design and build' with those of 'management'. A single firm is appointed, after a selection process that perhaps includes some degree of competition on price (price should not usually be the main selection criterion), to design, manage and deliver a project. Construction work and design is generally tendered for by specialist trade designers or contractors.

Components

The components of the design and manage system are:

- establishing the need to build
- establishing the client's requirements

- selecting and inviting tenderers to bid
- the contractor or contractors preparing their proposals for management, design, time and cost
- evaluation and acceptance of a tender which becomes a contract
- management, design and construction of the works.

The client as before will need to have in-house skills or obtain them in order to:

- prepare his 'employer's requirements'
- carry out his responsibilities in the contract or devolve these to an 'employer's agent'.

Variations on design and manage

The common variations of design and manage are:

- *Contractor*
  A project design and management organisation designs and manages the work, generally for a fee, and delivers the project by employing works contractors as its subcontractors to design and/or construct.

- *Consultant*
  A project designer/manager is the client's agent who designs and manages the work, obtains subcontract tenders from works contractors who then each enter into a direct contract with the client.

Considerations/characteristics

- the design and manage system has not been widely used in building procurement except where specialist buildings, often for the process industry, are to be provided
- the same criteria given for considering the 'management' system generally would apply. In addition, in selecting organisations to tender, their particular design record in the area of expertise involved is likely to be significant (in a similar way that some 'design and build' companies may have specialist skills or service particular market sectors)
- a difference from 'design and build' is that the initial price is not known when commitment to a 'design and manage' organisation is made – as with the 'management' system the works contractors are selected and paid at prices agreed by the client and the design/management organisation

- variants of evaluating the design and management fees exist, as with pure 'management'
- a client needs to know his functional requirements very well and then be prepared to leave the design and management to the appointed contractor or consultant in an area of building that is quite specialised.

Design separate from construction: traditional (sometimes called 'lump sum' contracting)

The 'traditional' system has been used for centuries with the function of design being provided direct to the client, linked quite often with a role for the designer that may have appeared to be one of management of the construction process also. In practice the management of construction by the designer was not the designer's role under traditional procurement.

In the traditional system the client appoints consultants for design and for cost control, then generally after design has been taken to any one of a number of stages, a main contractor is appointed to carry out the construction work.

Components

The components of the traditional system, shown in Figure 4.2, are:

- establishing the need to build
- establishing the client's requirements
- appointing a design team
- evolving the design and cost control
- the client's acceptance of design for the scheme
- preparing tender documentation
- selecting and inviting tenderers to bid
- evaluation and acceptance of a tender which then becomes a contract
- construction of the works.

The client will need to have in-house skills or obtain them in order to:

- brief himself and his design team
- carry out his responsibilities in the contract.

Variations in traditional

The common variations of traditional are:

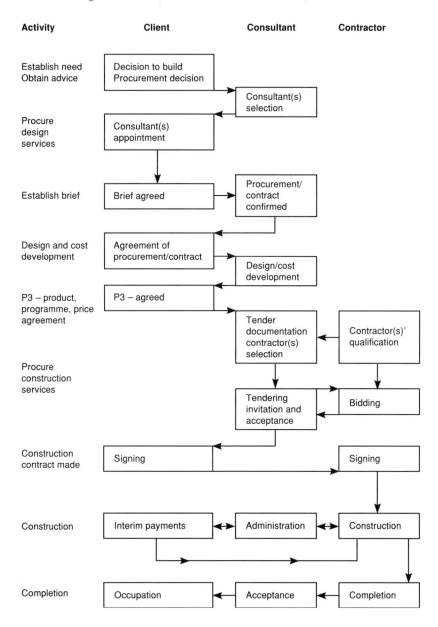

Figure 4.2   Components of the traditional system (sequential)

- *Sequential*
  A drawn and specified design is prepared and, together with cost documentation, contractors bid, generally in competition.

- *Accelerated*
  A contractor is appointed earlier in the sequence of design on the basis of partial information, either by negotiation or in competition. Negotiation, from the basis of the initial, partial information, takes place once the final design information becomes available. The components of this variation are shown in Figure 4.3.

- *Partial design of parts or elements of the works*
  A drawn and specified design is prepared by the client's design consultant and the parts designed by the contractor are then incorporated within the construction contract.

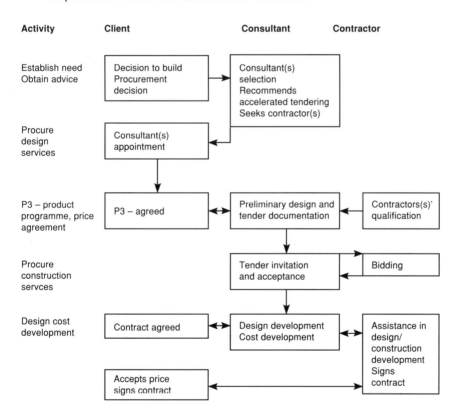

Remainder – as shown in the traditional system (sequential) Figure 4.2

Figure 4.3 Components of the traditional, two-stage system (accelerated traditional)

Considerations/characteristics

- the traditional system is very well-known, tried and tested in the UK (and in other parts of the world)
- experienced clients understand that if design has been fully developed at tender stage they should know their financial commitment before entering into a construction contract
- design can be carried out without undue pressures of programme or price because no contractor has yet been engaged
- consultants can be employed as and when the client wishes and/ or is advised to appoint them
- subcontractor or specialist design can be obtained in competition
- drawings and bills of quantities (nearly always used for contracts of a reasonable size) provide a common basis for tendering and evaluation is relatively easy
- design should ideally be fully developed at tender stage but variations on this can be adapted by accelerated, parallel activities of design overlapping with construction
- single or two-stage tendering may be appropriate; full design or tendering on approximate quantities are alternatives
- the overall period of design and construction, with design often being completed before construction tenders are invited, generally requires to be longer than is necessary for 'design and build' and for 'management' procurement options
- a longer period of overall design and construction may make the total project price higher because of increased periods of interim financing charges and interim payments to consultants and contractors
- cost control consultancy is required over a longer period before confirmation of estimated costs, verified by an accepted tender, is obtained
- if specialist subcontractors are involved the contractual responsibility of the main contractor may be diluted
- experienced clients have often found the system unsatisfactory for complex, large projects where certainty of completion on programme is a high priority
- separation of design teams from construction teams during the development of the project until tender stage may lead to the establishment of adversarial attitudes.

Design separate from construction: fee construction – the management method

The 'management' system has become widely recognised since the 1970s although a significant number of clients, consultants and contractors either have no or little experience of the system and its nuances. It is perhaps significant that the element of 'management' should have become separated as 'design' and 'construction' were already separated. The separation probably came about because the general perception of the construction process, and general reports on the building industry (see Chapter 3) bore this out, was that construction was an industry that was badly managed. Therefore a system that emphasised the management process and exposed and explored management expertise (which should have been there anyway but perhaps was hidden to the client and consultants in the traditional system) has come to the forefront of procurement options for many projects.

'Management contracting' and 'construction management' seem to bring out some emotion and feelings run close to the surface in those who are either for or against the system, more frequently so than in advocates for or against design and build or traditional systems. Management is, but need not be, an emotive subject.

One thing that should not be in dispute, nor should it really be a surprise although it seems to be, is that management projects are generally completed in a shorter time than traditional-route projects. The overall design and construction period is usually shorter and the construction period alone is often shorter. If this were not the outcome of putting more 'management resource' on to a contract it would be surprising and the concept of management contracting would by now have become fatally flawed. Comparisons of the extra costs this approach may or may not involve have been hard to establish but most independent comparisons are that 'management' costs more – how much more remains a question.

'Management' is a term meaning different things to different people. Variations in the system have been developed and more may come.

From what was perhaps its pinnacle of popularity, around the mid to late 1980s, management contracting was largely replaced by construction management in the early 1990s. I visited two well-known organisations to explore reasons for this and to find out modern developments in 'fee management' contracting.

Greycoat plc is a major property development and investment company with experience of all forms of building procurement. They have experienced in-house project management skills and are concerned to control the process of design and construction, rather than be an 'arm's length' client. They use construction management when they consider it an appropriate route and have introduced variations to it. For instance, they sometimes seek to obtain a 'guaranteed maximum price' from their construction manager and their trade contractors after the procurement process has reached a point at which the element of 'risk' can be negotiated to pass from the client to the construction team.

This will vary according to the circumstances of each job but will probably occur once the major elements of work have been tendered, but maybe not all let, and when the constructors are fairly confident that the client and its design team have effectively 'signed off' the design. Naturally there will be provisions for adjustment to any guaranteed maximum price if circumstances of change occur that warrant adjustment by client or construction manager. Another variation is that once the design has been finalised and most of the packages let the contract is converted to a lump sum design and build form and the construction manager takes on responsibility for the designer.

Laing Management discussed with me their provision of construction management generally, and in particular for a project of major prominence adjoining the Thames, in central London. The project has considerable constraints arising from the site, from the need to keep major infrastructure transport facilities operational whilst construction is carried out. At the stage at which I discussed construction management with Laing design had progressed for several years, with a team of first division consultants. A considerable amount of pre-ordering of materials and specialist design has been required on the project, over a relatively long time span, together with continuous cost control provided by a separate quantity surveying consultancy.

It is apparent that construction management has provided the client in the instances that I explored, and others generally throughout the late 1980s and early 1990s, with the capability to control the design and construction process on projects where a 'fee construction' procurement route is appropriate, as set out in more detail below.

It is also true that a number of clients have announced their intention to move away from construction management, particularly along the route of obtaining some form of price guarantee, it is said, 'to prevent spiralling costs'. To some this contradicts the whole philosophy of construction management. Another effect of seeking 'price guarantees' is that probably pure consultancy firms are unable to take on construction management work because their covenants are not secure enough. One construction management consultant has stated: 'it is disappointing that pure professional consultancy has not gathered more momentum. For a company of this size a covenant does not mean very much.' Another industry commentator has stated: 'enlightened clients want the best of both worlds, asking for a guaranteed maximum price or 'lump sum' prior to work starting on site'. To balance this view one of the industry's pioneers in construction management has said: 'if you are going for guaranteed maximum price once the design has been set and the packages let, then you are paying the contractor to take a risk that does not exist'.

It is apparent that procurement theory and practices are not set in stone.

Particular differences between management contracting and construction management are brought out under 'Variations on management' and 'Considerations/characteristics' below in revisions to the original text for this edition.

Components

The components of the management system, shown in Figure 4.4, are:

- establishing the need to build
- establishing the client's requirements
- possibility of appointing a design team before, at the same time, or after the appointment of, a construction organisation
- selecting and appointing a management organisation
- evolution of programme and design requirements, largely in parallel
- construction work itself is carried out by 'works contractors' in competition – the management contractor or construction manager usually does not carry out any of the permanent works although he usually provides temporary works, site hutting, safety and welfare facilities, site management personnel
- tendering, evaluation and appointment of perhaps sixty to a hundred works contractors
- construction of the works.

The client will need to have in-house skills or to obtain them in order to:

- brief himself, his design team and management contractor/construction manager
- carry out his responsibilities in the contract
- participate to whatever degree he wishes in the approval of options that occur throughout the management of the design and construction process. This interactive involvement by the client is inherent in the management system to a degree that is either not necessary or not possible in either the design and build or traditional routes.

Variations on management

The common variations of management are:

- *Management contracting*
  The appointed management contractor provides the service of managing for a fee all the works contractors who are to deliver the project by employing them as his subcontractors.

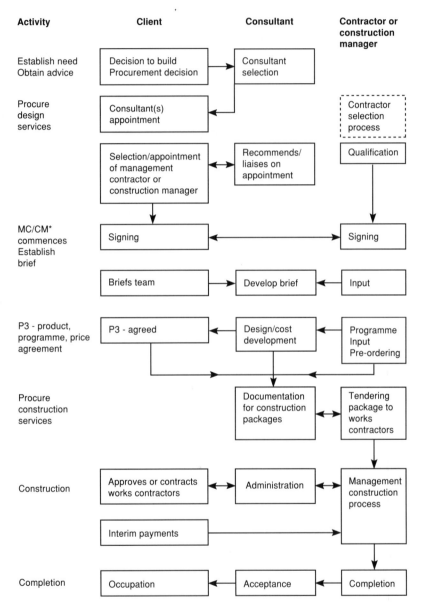

Figure 4.4   Components of the management system

It has been said that management contracting has the disadvantage that a management contractor is neither a traditional contractor, with the usual risks, nor a consultant (as a construction manager would claim to be) with equal status with a design team.

- *Construction management*
  The appointed construction management organisation provides the service of managing for a fee all the works contractors who are to deliver the project but each of them enters into a direct contract with the client. Construction management was relatively new to the UK in the 1980s. It originated in the USA and has been well described in its operation in that context in *Skyscraper, the Making of a Building*, Karl Sabbagh, 1989, Macmillan, and also in the Channel 4 television series of the same name.

  It has been said that construction management has emerged as a method in which the construction manager, by taking a more active role, can better manage the process. A construction manager could also provide cost control services (as generally provided by a construction manager in the USA) and even design services, providing then virtually 'design and manage' procurement. Construction management now appears to be preferred by clients that have the confidence and capability to follow a management procurement path. In addition to general contractors, who have offered a construction management service from their specialised divisions for quite some time, some consultants now specialise solely in construction management and more and more general consultants offer construction management services.

- *Fee management (prime cost contracting)*
  The appointed contractor provides the service of managing for a fee all construction work which is then carried out on a 'prime cost' basis, including the employment of subcontractors, to deliver the project. ('Prime cost' is discussed further in Chapter 6). Fee management is not considered as pure 'management' because the contractor does himself provide a considerable proportion of the labour, materials and plant and is often heavily involved in the doing, as well as the managing, of the works. It is considered that this is incompatible with the philosophy of a contractor supplying pure management expertise.

Considerations/characteristics

- The management system has become associated with projects that have some or all of the following characteristics:

- design is carried out by a design team independent of the contractor
- early completion is required
- the project is fairly large
- the project requirements are complex
- design and construction periods must overlap
- the project may entail changing the client's requirements during the construction period
- whilst early project completion is required maximum price competition for the works elements is wanted
- the contractor's or manager's responsibility for programme and eventual price is perhaps difficult to establish because the project details are of necessity not known when his appointment is made, therefore his responsibility can only be general, not specific in essential detail
- variants of evaluating the management fee exist, ranging from a percentage of final price, a scale of percentages of price, a lump sum fee, facilities for a lump sum fee plus a shared fee on the outcome of the service/contract goals that are or are not achieved. The apportionment of risk, reward and penalty between a management organisation and a client remains a subject of differing opinions in the industry and 'fee-sharing' is a philosophy easy in theory but difficult in practice
- some experienced clients have continued to use the system for their major projects, principally where programme is their priority
- attitudinal change may often be required by clients and consultants unfamiliar with nuances of the system
- it is claimed that 'them and us' confrontational attitudes change (and need to change) – communication improves because the contractor/manager is working for a fee, is more identified with the client and has a more limited risk than under either design and build or traditional routes
- the earlier appointment of the contractor/manager allows the contribution of his construction and management skills
- the design team very often feels that there is less time than appropriate to develop the design than when using the traditional plan of work programme that is usually adopted for a fully designed project
- when the client commences construction his certainty of final price is relatively low but his commitment to continue with the project is usually relatively high
- cost/price control of the project is often accorded a lower priority than programme, and perhaps quality, and very often becomes less easy to maintain

- attempts to provide so-called 'guaranteed maximum price' arrangements often fail on definitions and timing of what the offer is about and at what point in the programme it is obtained
- reduced project times are generally possible by continual evaluation of options and reallocation of resources to ensure that programme remains the priority
- quality and cost control often appear, by necessity, to have lower priorities than programme
- flexibility during design and construction may allow variations to be introduced more easily, but often more expensively, than under the traditional system
- subcontract tenders provide evidence of being obtained competitively, by being invited when the work is actually required, rather than invited months or years ahead of construction. The benefit of this competition is clearly passed on to the client
- firm price tenders for works contracts can very often be obtained because the work of a works contractor is often of a relatively short time span
- decisions on specialist design subcontractors are made jointly by the client, design team and management contractor/construction manager.

- The particular advantages claimed for construction management over management contracting are:

  - *transparency* – a construction manager is in a more 'honest' role than a management contractor. A construction manager is a consultant whereas a management contractor is also a contractor. Criticism has been made by sections of the industry that hidden cash discounts are given to some construction managers but reputable organisations deny that this practice is prevalent
  - *simplicity* – a management service is provided for a fee
  - *enforceability* – there is no intermediary between the construction manager and the works contractors, as there is in management contracting. It is claimed that this leads to fewer legal problems
  - *termination* – it is easier for a client to terminate the employment of a works contractor under construction management than under management contracting, where the management contractor is in contract with each works contractor. Also, if it becomes necessary to terminate the employment of the construction manager, that has fewer ramifications than termination of employment of a management contractor, principally because

a management contractor will have been in contract with the works contractors

- *payment* – a direct, therefore quicker, payment route between client and the works contractors makes for more control by the client
- *participation* – the construction manager and client can work more directly, with the construction manager becoming involved much earlier, if required, to advise on advance procurement of specialist works
- *flexibility* – any form of works contract can be used rather than the one preferred by a management contractor
- *professional roles* – a construction manager can provide all services of the other professions, if required, more easily than can a management contractor. If this takes place 'design and manage' would seem to have occurred rather than construction management.

Frequency of use of procurement options

As stated in the first edition of this book facts on the use of the different procurement options are not widespread but conjecture is.

At a time of much reduced workload in the industry it is understandable that some changes have occurred from the late 1980s and early 1990s when the first edition was published. Previously I referred to a survey carried out by the Royal Institution of Chartered Surveyors, Junior Organisation, in an attempt to discover facts on the use of different forms of building contract and therefore of different procurement methods. I have referred below, in my conclusions, to a survey of building contracts in use in 1995, carried out for the RICS by Davis, Langdon and Everest, a practice of chartered quantity surveyors and published in 1996. Around 900 questionnaires were sent out by the authors to quantity surveyors in private practice, local authorities and in central government and there was a response of around 22 per cent. It is suggested by the authors of this survey that it captured almost 15 per cent of new construction orders placed in 1995. I have also distilled my conclusions from information published in *Building* and similar trade journals on their findings of the use of different procurement options. My conclusions exclude overseas work, civil engineering, heavy engineering, term contracts, maintenance and repair work.

It is reasonable to conclude as follows:

- *design and build*, particularly through the variation of 'develop and construct', has become a very popular method, possibly accounting in the mid-1990s for around 30 to 35 per cent by value (say 12 to 15 per cent by number of contracts used) of building procurement, which is an increase of three to four times since the late 1980s. This may be because there is a standard JCT form for use with design and build or develop and construct (unlike for construction management) but it is much more likely to be because of the 'one-point responsibility' that clients often like to obtain and because of the ability to progress a scheme so far with a 'scope designer', for a relatively small outlay in professional fees. Of course there are hidden design costs that a client may think he has avoided by using design and build or its variants, but he should realise that successful contractors do recover professional fees somewhere. The practice of 'novation' of the client's designer to the design and build contractor remains very popular with some clients. Sometimes a client will also retain his designer whilst at the same time requiring novation of him to his contractor. A conflict of interest may well arise from this arrangement and apparent savings in fees may well be outweighed by having 'one line of design responsibility' but with some possibility of conflict existing between client and contractor for the designer. In the mid-1980s design and build accounted for around 15 per cent by value of building procurement.

From RICS surveys it is apparent that the JCT Contractor Designed Portion Supplement was used on perhaps around 30 per cent of 'traditional' contracts and so even in this area design has been moving, at least in part, towards the contractor.

- *traditional* has maintained much of its popularity in use since the late 1980s, and probably still accounts for over 50 per cent by value (perhaps over 80 per cent by number, but this includes over 25 per cent for the JCT Minor Works form) of procurement, outside of industrial work. During the 1990s, when workload has been low and the urgency of building has largely gone, competitive tendering on so-called 'lump sum' methods has remained fairly popular and constant. JCT 80 appears to account for around 33 per cent by value (20 per cent by number) of all building procurement. In the mid-1980s traditional accounted for around 75 per cent by value of procurement
- *fee management* currently has a market share of, at best, around 10 per cent (by value) with construction management fluctuating in popularity with management contracting. This figure is only obtained because the size of the projects on which management is used is, on average, much larger than for other procurement routes. In terms of numbers of contracts let using a fee management route this probably accounts, in 1996, for less than 1 per cent. In the early 1990s construction management had risen to become more

popular, perhaps accounting for around 15 per cent by value of the total market at that time but it has now (in 1996) dropped back to less than 5 per cent by value. The number of larger-scale, more complex, projects decreased in the 1990s, principally because many of the 'major players', such as speculative developers, had then withdrawn from their activities. Also management contracting had suffered some unfavourable comments at the end of the 1980s and now it seems the time for construction management to be suffering similar, less favourable 'press'. More consultants, as well as contractors, now offer construction management and some clients seem to prefer this but if 'guarantees' are required then consultants may be at a disadvantage relative to contractors.

In summary, it seems fair to say that traditional is remaining the most used procurement route, but it is being pressed more than ever by design and build. Management continues to fluctuate according to the number of clients and schemes that require that form of procurement. The above also demonstrates that fashion has some influence on procurement choice.

Summary

Design combined with construction – design and build and design and manage – should be considered when:

- a building is functional rather than prestigious
- a building is simple rather than complex, is not highly serviced and does not require technical innovation
- a brief for scope design is unlikely to change
- a firm price is needed in advance of starting construction, although with design and manage this is not achieved
- a programme can be accelerated by overlapping design and construction
- a single organisation is required to take responsibility and risk for design and construction.

Design separate from construction – traditional (or 'lump sum') – should be considered when:

- a programme allows sufficient time
- consultant design is wanted
- a client wishes to appoint designers and constructors separately
- price certainty is wanted before the start of construction

- product quality is wanted
- a balance of risk is to be placed between the client and constructor.

Design separate from construction – management – should be considered when:

- an early start to construction and an early programme completion, requiring design and construction to proceed in parallel, is wanted
- flexibility in design is wanted to allow for changes to be made as the process of design and construction is carried out
- a project by its nature is organisationally complex, probably with a need to manage a multiplicity of client, consultant and contractor organisations
- a project is technologically complex resulting from often differing requirements of future users
- maximum price competition for the works elements is wanted
- a client and his advisers have sufficient management resources as demanded by the interactive nature of this procurement route.

5 Choosing the Route

The procurement route that is appropriate to the overall balance of objectives and to client priorities for each project should arise from those objectives and priorities. The marketing skills, promotional activities, advertising claims of organisations which suggest they are unique or appropriate to nearly all procurement paths should be examined closely. Some of the claims may be quite valid, if occasionally perhaps overstated, but arguably no organisation offers services which are unique – at least not generally for very long. The evolution of variations of an old theme, as though they were new, unique themes, will no doubt continue as part of the ways apparently necessary in current promotional presentation to beat the competition. An example of this is 'partnering' which is currently receiving more attention and will probably soon be extolled as the answer to nearly all procurement needs, before it is then seen in the context of other options.

In this chapter the procurement assessment criteria, abbreviated to PAC, that are used for choosing the procurement route are first described and assessed. Then the procurement arrangement options, abbreviated to PAO, are given as they arise from assessment of each of the PAC.

The assessment of which route to choose is made from the client's point of view. Then considerations and characteristics of each route in relation to contractors and consultants are given and the merits and demerits of each system are stated.

Case studies are given in Chapter 7 to illustrate the use of each procurement route.

Procurement assessment criteria (PAC)

The enterprise of construction involves risk at nearly every turn, in almost every decision that is made. Once organisations, both consultancies and contractors, have been employed, then timely, correct decisions are essential to lessen the risk and eventual costs to a client and, not inconsequentially, to the consultants and contractors themselves.

The design and construction of buildings is a balance, a compromise in the circumstances existing at the time between *quality*, *time* and *cost* or, put alliteratively, *product*, *programme* and *price*. Each one of

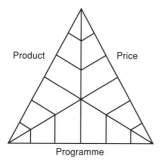

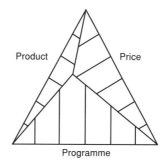

Product / Price

Programme

Equal balance of priorities
between product, programme
and price

Product / Price

Programme

Programme has priority with
price second and product third

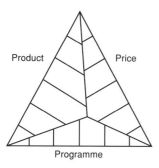

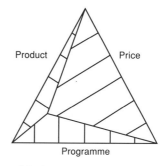

Product / Price

Programme

Product has priority with price
a close second and programme
third

Product / Price

Programme

Price has priority with
programme second and product
third

Figure 5.1   The balance of product, programme and price

these three constraints most probably, not always, pulls against one or
both of the other two. Invariably product quality will cost more and
may take longer to achieve; programme time may need to be shorter
than thought to be ideal, perhaps for commercial reasons, thereby causing
product quality perhaps to be compromised and the price to increase,
although hopefully these possible disadvantages will be outweighed by
the overall value of a shorter programme; price may be the determi-
nant of the project proceeding or stopping.

Achieving the balance of product/quality, with programme/time and
price/cost is the challenge of building procurement. Figure 5.1 shows
the way the balance changes from project to project.

Identifying your priorities

As a construction industry customer, you can choose from a wide selection of procurement routes. Which one suits your business needs?

To help you home in on a suitable procurement route, the following customer priority checklist has been prepared as a prompt for discussions with your principal adviser. Study the list of project priorities, A–1. Consider each in turn and decide which option (1–23) most accurately reflects your preference. Move along the line and note where each procurement route gets a positive score. When every priority has been marked, you may simply add the scores in each column to give comparative totals.

Alternatively, you may wish to refine further the initial broad assessment by giving extra weight to certain priorities. If getting a firm price, say, is measurably more important to you as a business customer than getting an early completion, you could weight the scores accordingly. Priority E might be worth a score of 3, for example, to Priority A's 1.

A Timing	How important is early completion to the success of your project?
B Controllable variation	Do you foresee the need to alter the project in any way once it has begun on site, for example to update machinery layouts?
C Complexity	Does your building (as distinct from what goes in it) need to be technically advanced or highly serviced?
D Quality level	What level of quality do you seek in the design and workmanship?
E Price certainty	Do you need to have a firm price for the project construction before you can commit it to proceed?
F Competition	Do you need to choose your construction team by price competition?
G Management	Can you manage separate consultancies and contractors, or do you want just one firm to be responsible after the briefing stage?
H Accountability	Do you want direct professional accountability to you from the designers and cost consultants?
I Risk avoidance	Do you want to pay someone to take the risk of cost and time slippage from you?

Figure 5.2 Identifying your priorities. (Reproduced from pages 6 and 7 of the BRT report Thinking about Building, *with the permission of The Business Round Table Ltd)*

- identify his ability to take risks and assess the effect that the occurrence of each risk (for example, delay to the completion date or increase in price of the product) would have on his business
- identify risks that should be avoided if at all possible (for example, delay to completion date)
- formulate a policy for managing risks (for example, the results of fire) that may not or cannot be avoided.

Risk can be found (i) within an organisation and (ii) outside that organisation, for instance arising from the effects of employing another to carry out work. It is important to realise that risk cannot be devolved and remains with a person or organisation until responsibility for the effects of that risk has been moved to and accepted by another. Risk will generally increase, both for a client and also for other organisations, when other persons have been employed by a client. It is important for a client to know how, and to what degree, risk has consciously been passed to another organisation, or how it has been shared in some proportion between him and another, how the risk may not have passed on at all, or indeed how the risk to him may have been increased by the employment of another.

Contemporary practice on the placement of risk varies but good practice is to allocate risk between parties to a contract on a rational basis so that each risk is taken by the party best able to assess, carry and manage the risk. If risk is transferred or spread between a number of organisations it is important that the risk-takers have the ability to take all the risk otherwise the risk-placers will not be fully protected. Because many organisations in building are small companies it may not be possible for them to take on some risks due to their insufficient capacity, arising from limitation of company status or through an inability or unwillingness to insure.

In order to assess risk avoidance it is necessary to identify what risk is. Risk is discussed in more detail in Chapter 6 where 'fundamental risk', 'pure and particular risk' and 'speculative risk' are explained. In simple terms 'fundamental' is normally non-insurable and is the risk of events such as war happening. Construction contracts are not usually drafted to require a party to include a price to cover 'fundamentals' and as such the risk will lie with each party to suffer the consequences. 'Pure and particular' risks are of collapse, storm, tempest, subsidence, injury to persons or property and are usually carried by the contractor, who often is required to insure for the effect of such risks upon the works. 'Speculative' risks are those that can be varied in the incidence of their effects between the parties, such as bad weather, industrial action, lockouts, effects of variations on time, discrepancies in tender documents and so on.

In relation to the three procurement routes described in this book and the contracts usually available for them, the risks of failure to product, programme and price are assessed very broadly as follows and Figure 5.3 illustrates, again in broad terms, where the balance of risk is likely to fall in each procurement method:

- *Design and build procurement*
    - product – the quality may be as offered by a contractor or, more probably, as specified to a smaller or larger degree by a client's team and possibly developed by a contractor
    - programme – the contract period is fixed but will be subject to extension if the client/designer introduce changes, therefore some risk of over-run
    - price – the contract price is fixed but is subject to adjustment if changes are introduced by the client/designer, therefore some risk of price increase.
- *Traditional procurement*
    - product – the quality is as set out by the client's team and with proper monitoring of the contract requirements the quality level specified should be obtained
    - programme – the contract period is fixed but will be subject to extension if the client/designer introduces changes, therefore some risk of over-run
    - price – the contract price is fixed but is subject to adjustment if changes are introduced by the client/designer, therefore some risk of price increase.
- *Management procurement*
    - product – the quality is as the client's team specifies and with proper monitoring this should be obtained
    - programme – whilst a programme is not usually a contract term a target date will be the objective of the management contractor or the construction manager but there is a risk of over-run
    - price – price will not usually be fixed before construction starts, therefore a risk of price increase.

Another way of looking, very broadly, at risk and its avoidance across the procurement options is to consider the *nature* of the client, the *nature* of the project and the *complexity* of its construction as follows:

Standard construction/relatively simple end product
Procurement route of design/develop/construct (any variation such as 'develop and construct' as is appropriate to the circumstances of the project) is probably most suitable by giving single-point responsibility to the client who should have whatever initial design/planning advice he needs before going to construction tender, an employer's agent to

liaise with the appointed contractor who takes on virtually all the pure, particular and speculative risks.

*More complicated, more detailed construction; more specific/
complicated requirements of client in briefing stages*
Procurement route of traditional, so called 'lump sum', probably is the most suitable (any variation of traditional as is appropriate to the circumstances of the project), so obtaining from the industry what it is most familiar in providing, *provided* efficient pre-planning/design, documentation and contract administration are carried out. Risks are fairly balanced between client (who probably takes on the speculative risks) and contractors (who probably take on the pure and particular risks), *provided* design has been properly done before going to tender.

*Innovative/complex design and construction; hands-on client capable
and desirous of strong leadership*
Procurement route of construction management is probably most suitable to allow specialists and trade contractors to contribute on design and buildability economies at crucial stages in the process. Management routes suit larger projects but need not be restricted to them. Risks are largely left with the client even if some form of GMP is introduced at a later stage in development of the design because, by then, there is much less risk involved.

It is theoretically possible to have all the risks of design, management and time slippage or error put into the hands of one organisation. With an excellent 'employer's requirements' answered by an excellent 'contractor's proposals' a design and build contract could possibly pass nearly all the usual risks, at a price, to one organisation, to a single contractor. However in practice it has not been so easy to achieve. One of the criticisms made of some of the invitations for the Private Finance Initiative (PFI) has been that attempts have been made to put far too many, if not all, of the risks on to the contractor. Very onerous terms have been demanded and this has resulted in relatively few organisations wishing to tender. It is not known what effect very onerous terms and the acceptance of high risk by contractors has had on the price levels of tenders.

It is not possible to say in any hard and fast way that one procurement route, as opposed to one type of contract versus another type of contract, has more risk than another procurement route, in all possible circumstances. It will depend on the balance of the project's priorities.

Figure 5.3 attempts to indicate where the balance of risk lies under each PAC, under each PAO. The balance of the eight PAC will show where risk, its avoidance and acceptance, lies. In relation to *Thinking about Building* three groupings for risk avoidance have been given.

- *A client prefers to retain control and therefore risk*
  Quite clearly under the 'management' forms of contract the risk of price and programme slippage remains largely with a client. Under 'traditional' the balance of risk is probably over 50 per cent with a contractor. Under 'design and build' the risk has moved predominantly to a contractor.
- *A client is prepared to share agreed risks*
  If this is a client's decision the traditional ('lump sum') and management methods allow risk-sharing whereas 'design and build' is not structured to make this so appropriate.
- *A client wishes to place (and pay for) the risk of cost and time slippage*
  'Design and build' is ideal to meet these criteria.

Procurement Assessment Criteria (PAC)	Procurement Arrangement Option (PAO)					
	Design and build		Traditional		Management	
	Balance of risk		Balance of risk		Balance of risk	
	Client	Contractor	Client	Contractor	Client	Contractor
1. Programme (timing)	1	4	2½	2½	1	4
2. Variation	1	4	2½	2½	4	1
3. Complexity	1	4	4	1	4	1
4. Product (quality)	3	2	1	4	1	4
5. Price	1	4	1	4	4	1
6. Competition	1	4	2½	2½	4	1
7. Management	1	4	2½	2½	4	1
8. Accountability	2	3	4	1	4	1
9. Risk (overall)	1	4	2	3	4	1

Scale: units of 1 to 5

Figure 5.3   Assessment of risk between the procurement routes

Timing (programme)

Timing or programme is arguably the principal reason why the traditional procurement route has lost some of its previous preeminence to procurement by design and build or by the management routes. Since 1980 a number of clients, particularly those involved in large and therefore often prestigious projects, perhaps with a profile known and seen by the public, have concentrated on procurement methods where an overall fast programme time was high on the client's priorities. Examples of such projects have been air terminals, City of London office developments, shopping developments, an international conference centre that needed to be completed on time because it was required for a known date, when heads of state were to attempt to advance the affairs of nations. The Olympic Games, the World Cup or whatever cannot take place if associated construction is completed too late. Programme (another word for timing) has often become predominant.

There are two aspects to programme and timing. One is completion at the time, on the exact date, expected by the client when he enters into discussions and then into contracts with his consultants and/or contractor(s). The second aspect of timing is in the sense of early completion, that is, earlier than would otherwise be. The jargon of the era in building calls early completion 'fast track', that is, completing a project quicker than by traditional means. The NEDO building reports *Faster Building for Industry*, 1983, and *Faster building for Commerce*, 1988, have repeatedly drawn conclusions that the majority of buildings are constructed too slowly in the UK, in comparison both with overseas experience and also with the best examples in the UK where construction projects have been built quickly.

If overall programme time is crucial to a client he and/or his principal adviser should be absolutely clear why this is so. The placing of programme criteria very high, perhaps as the highest, in the client's priorities must be valid and should not be as the result of an unchallenged client assumption or one which has simply arisen as the result of an undisciplined client who could not make up his mind to proceed with the project earlier. This may indeed be the case but it is wise to recognise it and then to accept the position rather than take 'fast track' as an immoveable, unchallengeable 'given' in the procurement criteria. Programme will often, for whatever reason, be a priority for some clients but they should appreciate, or if not they should have it explained to them, that programme very often has a premium both in price and quality of product.

The other aspect of programme is the degree of certainty or otherwise under the different procurement routes, and therefore in consultant and construction contracts, that will apply in each route. Certainty of programme, that is, completion, is not guaranteed until a client's brief, followed by a scheme design, is completed and a construction contract has been signed. Even then the exact contract conditions may make completion to programme less certain than a client unfamiliar with building may assume.

If programme timing is regarded as 'crucial' the following procurement arrangement options (PAO) will generally be appropriate, in the following order:

   (i) design and build (probably by negotiation rather than by competition, for a relatively simple building)

  (ii) management contracting or construction management (the building needs to be of ample size and complexity in order to make the relative extra price of this system worthwhile – benefits in programme should be assessed in relation to any extra price in construction costs, principally by devoting more resources to 'management')

 (iii) accelerated traditional (probably by negotiation, not usually by competition although a two-stage tender may be appropriate)

 (iv) design and manage, say by BOOT

  (v) traditional (if all design is complete at tender stage and variations are not introduced thereafter – however the overall length of programme time of sequential design and construction combined probably cannot satisfy using this route when 'crucial' programme criteria exist).

The choice from the PAO will arise out of the exact circumstances of the client and the project. Where programme timing is 'crucial', considerations of design quality, price and maybe competition and price certainty at the start of the job will have to come much lower in procurement assessment criteria (PAC).

If timing is regarded as 'important', the following PAO will generally be appropriate, in the following order:

   (i) design and build (probably by negotiation rather than by competition, for a relatively simple building)

  (ii) management contracting or construction management (comments as above under 'crucial')

 (iii) accelerated traditional (ditto)

 (iv) design and manage

  (v) traditional (ditto).

Where timing is 'important,' considerations of price, quality and price certainty will be lower in PAC.

If timing is regarded as 'not as important as other factors' the follow-ing PAO will generally be appropriate, in the following order:

(i) traditional (the overall programme time for design and construc-tion is generally longer than in other procurement routes but most of the other PAC can be met more easily by this route)
(ii) design and build (if programme time is not important and client brief/design difficulty/complexity of building are also not involved, the single-point responsibility of design and build may become a significant PAC).

Where timing is 'relatively unimportant', price certainty, competition and more certainty over responsibility and product quality can more easily be achieved and the PAC for these weighted accordingly.

Controllable variation

'Variation', or the newer, transatlantic term that has come into use, 'change order', is undoubtedly among the most emotionally charged concepts in the construction industry.

Much of life is subject to change of mind, rethinks, reappraisals and so on but variations in building contracts very often produce divisions and conflict between client, consultants and contractors. This has led to revisions to standard forms of contract in an attempt to maintain a balance of risk and remedy within construction contracts between cli-ents and contractors. It has also led to an industry of seminars on ascertainment of loss and expense settlements and to an army of con-struction contract consultants. The findings of the Latham Report were that the industry was 'too confrontational', largely in my opinion re-sulting from differing interpretations of the effect of 'changes', that is from 'variations' introduced during design, and, much more serious, during the construction phase.

New procurement routes have arisen in large measure because of these dissatisfactions, in an attempt to avoid the pain and displeasure of reaction to 'variations', particularly when projects often cannot be designed and tendered in such a way that variations can be avoided.

To understand the concept of 'controllable variation' it is worth under-standing the need for variation, controlled or uncontrolled, at all. Why, for instance, if the RIBA *Plan of Work* is followed reasonably strictly do variations occur? Leaving aside design and build, management and fast-track procurement routes in general, why do variations occur on projects, particularly during the construction stages, when the theory of the design process is that by tender stage design should usually be complete? It is necessary to understand the nature of the design process and to see why 'variations' should occur and then to understand the importance, if at all, to the client of being able to introduce variations.

The RIBA *Plan of Work* is very well-known in the construction industry but it is probably true to say that it is followed strictly only on the minority of projects, even on those following the traditional procurement route. The RIBA *Plan of Work* is a sequential progression through the stages of design and construction of a building project. It assumes the client appoints an architect and that the project passes through the 'inception' and 'feasibility' stages, proceeds through levels of increasing detail in 'outline proposals', 'scheme design', and then into 'detail design', 'production information' and 'bills of quantities', 'tender action' and then the stages of construction and completion on site. It is an iterative process (not one of sequential movement) of going through cycles, and sometimes around in circles, along paths of progressively increasing detail that are necessary in order to move from broad concept to detailed design reality.

The theory is that at tender stage 'all information is complete in sufficient detail to enable a contractor to prepare a tender'. The practice, rather than the theory, is that the information is indeed quite often in sufficient detail to enable a contractor to prepare a tender but is that information sufficient to enable construction to be carried out without the supply, at the right time, of significant other information to the contractor?

The theory of the standard forms of building contract for traditional contracting is that the building has been fully designed (this is also the fundamental philosophy and thrust behind the recent introduction of *Coordinated Project Information, the Standard Method of Measurement of Building Works, Seventh Edition, the Common Arrangement of Work Sections for Building Works, and supporting Codes of Procedures for Production Drawings and Project Specification, and Code of Practice for Measurement*).

Variations, controllable or not, may have to occur if the tender/contract information is insufficient/incorrect. Contract information may have to be modified by the issue of variations arising principally from the design team, although the variations may in some instances lead back to the client's own earlier inability to have resolved some of the issues that, once resolved, result in variations. The other major source of variation is change introduced solely by the client once the building is on site. It may be that invention or the latest technology requires modification of an industrial or commercial process and the building for that process must therefore be modified from that which the contractor has contracted to build.

With all the goodwill, or otherwise, of dedicated and less dedicated clients and consultants it has been the author's experience, over many years, that a very small proportion of new construction, let alone alteration works, is tendered and built without a significant number of variations. It is a fact. It may or may not be regrettable, it causes emotion to be generated, extensions to the contract time for the effect of variations to be sought, to be accepted or rejected in varying degrees, claims for losses and expenses incurred in claimed disturbance of construction programmes to be made, accepted, rejected, arbitrated, litigated and so on – a minor confrontational industry in itself. Hence moves by Latham to have 'fast-track adjudication', which has now been given statutory backing.

Some clients have tried to control the design process to avoid unnecessary variations. Significant public clients, used to building regularly, such as the PSA and Department of Health, had their own equivalents of the RIBA *Plan of Work*. The PSA had its *Plan of Work* and the Department of Health its *Capricode* system. With the implementation of such systems, in some instances the concept of 'a certificate of readiness to proceed to tender' has been introduced as a discipline on the design team. The certificate of readiness is a statement by the design team that their design is in fact complete and no significant variations will arise due to incomplete tender design information, that will require a variation to the contract to be issued in order to obtain the required construction.

'Controllable variation' is perhaps therefore the key rather than numerous variations. It is fruitless to continue to expect a variationless contract and the traditional procurement route has recognised this (a) by having such a significant number of contract conditions that provide for the issue and valuation of variations within a traditional, 'lump sum contract', (b) by including within the contract conditions provisions for dealing with disturbance of a building programme and (c) by introducing forms such as the JCT Standard Form with approximate quantities. Further, some clients, aided by some of the major contractors, knowing that variations will occur, have sought to avoid the uncertainties of the result of introducing them by using a management procurement route. Alternatively they have tried to avoid 'designer-inspired variations' by using the design and build categories of procurement.

The client should be aware and be advised of the effect of uncontrolled variation, particularly during the construction stage. Variations emanating from the client (not those that are the regrettable result of incomplete design and specification by consultants at tender stage) must be assessed and the effect of them drawn to his attention, including taking into account the probable effect on programme, product and price. Advice from consultants and contractors should be obtained before variations are introduced. This is the correct way to consider and to introduce variations of any significance, in a controlled way.

If a client is unable to fix his brief and he has a requirement for 'controllable variations' high on his list of priorities the following procurement arrangement options (PAO) will generally be appropriate, in the following order:

(i) management (the very flexibility of the management route allows variations to be introduced within a framework of less certain contractual conditions)

(ii) traditional (well-known rules are contained in most contracts for ways of dealing with the possible effects on programme time and the possible price effects of introducing variations, but uncertainties remain of knowing, in advance, the exact effects of significant variations)

(iii) design and build (because of the single-point responsibility contract the client may be less able to know or to negotiate effectively on the programme, price and quality effect claimed for the introduction of a variation)

(iv) design and manage (as above).

It is emphasised that variations very often pull in the opposite direction to the PAC of programme and early completion. They also often can disrupt the normal progress of construction events and can cause the quality of the finished building to suffer.

If controllable variation is 'definitely not' expected on a project the following PAO will generally be appropriate, in the following order:

(i) design and build or design and manage (taking into account other PAC, particularly those of complexity and quality)

(ii) traditional sequential (this route is after all what the traditional system is supposed to be about – full pre-design, no variations. PAC of timing and responsibility may be relevant and affect the choice or otherwise of this system).

Complexity

Complexity is important in deciding procurement because simple construction may best be purchased one way and complex design and construction another way. The complexity of the function of the building, that is, what goes on in the building, is not necessarily significant but the complexity of the construction itself and of the environmental services of the building are usually significant.

Construction can be divided broadly into two types. Division can be made, first, into buildings principally involving people such as houses, offices, shops, hospitals, some factories and warehouses, and, second, into those buildings required for industrial processes such as chemical plants, car-making plants and the like. Buildings for industrial/commercial processes may require to house some people within them but building function and its processes are the major determinants of the construction and environmental services of such buildings.

In our present society the large majority of buildings erected are for use by people rather than simply to house processes. Having said that, the increase of automation, information technology, robotics and so on constantly affect the balance between people's involvement and the attention that should be given to technology and machinery. The majority of buildings required are still relatively technically simple and require only simple environmental services. This is the case with much housing, offices, simple warehousing, small retail units and local hospitals and the like. However, some offices (those for instance with computer facilities and financial dealing 'desks'), more complex hospitals, clinics, enclosed shopping facilities, department stores and similar structures will require a high level of environmental servicing. Also, perhaps because of the size and visual impact of these structures on the awareness of the community, their design and construction will probably also be complex. To help in this area, 'environmental impact analysis' is a relatively new skill that is used to evaluate, in advance, the probable effect of major new projects on the environment.

Buildings for industrial/commercial processes may need to be technically advanced, truly 'high-tech' for such processes as electronics, 'micro-chip' production and pharmaceuticals. The building shell may need only be a relatively simple envelope but sophisticated environmental services, humidity control, air filtration and specialist electrical standards may be required.

Complexity as a PAC is not therefore something that can be generalised by building type – it requires to be assessed and it should become apparent early in the project's life.

If the building (as distinct from what goes into it) needs to be 'technically advanced and/or highly serviced':
Any of the procurement routes may be considered, except perhaps design and build, in conjunction with other PAC. Aesthetic requirements will help in deciding between traditional and design and manage procurement, programme criteria between traditional and management procurement. Design and build is unlikely to be appropriate for complex solutions, except perhaps in specialised sectors of building for industry. It is more likely that design and manage rather than design and build will be appropriate where complexity may involve a relatively long design period with client involvement throughout.

If complexity is only 'moderately present':
Any of the procurement routes may be considered, in conjunction with other PAC.

If the building is not complex:
In this case design and build should be considered. It will provide single-point responsibility and may have other benefits. Other PAC will probably then determine the final procurement choice.

Quality level

Quality is a subjective value but it need not always remain so. It is accepted that quality for one person is not necessarily the same as for another and for this reason specification of quality and setting a specification level is attempted. Ultimate, visual, external quality of architecture and building was thrust into the national consciousness and debate by the Prince of Wales during the period 1987–9. The Prince stated his ten 'principles of design', some of which deal with quality, in *A Vision of Britain*, 1988. It is not appropriate or necessary to comment here on the detail of his views. The debate on design has been useful because it has caused architects, developers, planners, buildings themselves, the built environment (in the clumsy vernacular), even society itself, to be more carefully thought about. Hopefully, this will remain so always. Buildings are generally too big, too costly and too long-lasting to be thrown up or cast down lightly. Now that proceeds of the National Lottery are going to the Arts Council and the Sports Council architecture is again entering the national debate.

If *quality* is, and perhaps must remain, largely subjective, *quality level* can hopefully be looked at by more objective standards. Quality level may be divided into two parts – quality of the materials and workmanship and quality of the design concept. The quality level of the materials and workmanship is discussed first.

At present the finished standard and quality level of buildings in the industry generally is often poor regardless of the procurement route adopted. In my meetings with the Construction Round Table (CRT) I discussed 'quality of product'. The CRT is committed to making significant improvements to the performance of the construction industry, including quality of the finished product. Members of CRT include companies such as British Airports Authority and Marks and Spencer, both committed to quality in their own businesses. Among the CRT aims of continuous improvement in quality level is to achieve in building a standard of a production item that is not a prototype.

In any procurement system the quality level, that is, the quality of the product, what is called in construction terms the specification level of materials and workmanship that is to be provided for the price to be paid, should be ascertainable and should be capable of judgement in advance. The quality level of the product offered should then be capable of being checked with the built product that has been provided. In discussing any procurement system it is unacceptable to say that one method will or will not give a better or poorer quality level as such than another system.

As stated above quality is subjective but quality levels can be made objective by specification and the specification can then be checked. Consistent with other constraints of programme and price a quality level should be set and it should be provided – otherwise offer and acceptance and the contract conditions have not been met.

In *Thinking about Building* three grades of quality level have been given – 'basic competence', 'good but not special' and 'prestige'. Because *TAB* of necessity uses brevity and shorthand it cannot properly convey the large, subtle ranges that exist between 'basic', 'good' and 'prestige'. Perhaps for this reason the matching of quality level, that is, specification level, with procurement routes is not too fruitful, bearing in mind the earlier comments. Many prejudices exist, for instance that design and build inherently cannot provide a quality level equal to that obtainable with traditional or management procurement. The suggestion of *TAB* is that if you want 'basic competence', design and build will give it – it should certainly be understood that the other systems also will, all too readily, provide 'basic competence'. Likewise the suggestion is made in *TAB* that design and build cannot easily give you 'prestige', but that 'good but not special' is obtainable with any procurement system. Before leaving this point it is emphasised that quality, referring to specification of materials and workmanship, should be attainable to whichever level is required if (a) the contract specification is sufficiently detailed and (b) mechanisms and procedures of inspection are adequate to ensure that the product supplied is equal to the specification tendered. The fact that too many design and build, traditional and management contracts have instances of poor materials and workmanship is not inherently because of the system used – it is the result of its poor operation and lack of monitoring.

The second aspect of 'quality level', namely the quality level of the design concept and/or of the design detail, is much more difficult and, some may say, is linked very closely to the quality level of specification. It is sometimes so but it has already been demonstrated there is a difference. Quality level of design begins to combine subjective and objective criteria. It should be relatively easy to assess the quality level of design that is provided under the procurement systems for relatively objective design matters such as function and economy. It is not so easy to assess relatively subjective design matters of form, commodity and delight. Quality then becomes opinion and so easily becomes debatable.

Before leaving quality level, in relation to design quality, it is generally likely that the procurement routes given below will be appropriate for the levels indicated in *TAB*.

For 'basic competence' and 'good but not special':
All procurement routes will be appropriate.

For 'prestige' quality:
It is most likely that only traditional and management procurement can provide prestige quality. This is simply because for 'prestige' (for example, the National Theatre, the Lloyds building in the City, the National Gallery, the Westminster Conference Centre and so on) the briefing process, that is, the period spent agreeing the function, the exact space standards and the interrelation of spaces, must involve the client in decisions on detailing of design and specification. It is most unlikely that this process can be obtained in the UK outside of the present arrangements of separate, private architectural and consultancy practice and separate contractors. 'Prestige' designers require a challenge that is currently only available to them, sufficiently often and in sufficient variety, by practising outside the ties of sole patronage or employment from a construction company. Architecture of the highest quality in the UK has only been produced that way in recent years. The future may prove otherwise but there are few signs at present of this coming about.

Price certainty

Price certainty is often a critical PAC.

Before the current range of procurement options became available, the traditional route, if operated properly, would generally provide price certainty at the point of letting a construction contract. If design had actually been completed, then a tender based on that design should therefore have given a good indication of the final contract price, provided variation was not introduced during the construction period. Inflation during the construction period would either have been included in the contract sum or otherwise adequately allowed for.

'Price' and 'certainty' need definition.

'Price' should be taken to mean the total constructional costs of design fees, construction contracts, financing costs and client management/monitoring costs.

Price as defined can mean that a construction contract worth £10 million can, in addition, often generate design costs, cost consultancy and site inspection/supervision costs of 15 to 20 per cent, financing costs of 5 to 15 per cent (depending on the programme time and on interest rates throughout the design and construction phases) and direct client costs of perhaps 1 to 3 per cent. So £10 million of construction contract value means perhaps £13 million of price, that is cost expenditure to the client. In addition to this there may be a developer's margin, associated agency advice, land costs, land purchase and sales costs and other related expenses.

'*Certainty*' should be taken to mean relative assurance, within defined parameters of time and money budgeting, to a factor of say plus or minus 10 per cent at the start of construction provided a scheme design and principal detailed design is complete. If this is not the case a plus or minus factor of between 10 and 20 per cent may be involved, depending on the project circumstances.

Certainty, in everyday life, is often just that – yes or no. In commercial enterprise certainty, meaning absolute assuredness, is seldom possible when talking of price, especially over the period of time that is involved in designing and constructing buildings. During say a three-year period many market forces, tender prices, interest rates, general inflation rates and related factors may change significantly such that price certainty must remain a relative term.

Price certainty will change, in the degree to which it is certain, during the life of a project. In the early stages the plus or minus factor will be greater and, as detailed development of the project continues, the risk factor included in the price certainty should become smaller. Figure 5.4 illustrates this change.

Before a procurement route is chosen discussions with a client should establish:

- how critical price certainty is to him
- what degree of risk can/should be allowed within that certainty
- how the client will want and/or need to monitor and adapt the process of design and construction if forecasts of price certainty are found not to be achievable.

As often with procurement decisions, other PAC such as programme and competition may influence the degree to which price certainty can be achieved. Because by far the major cost of building is only committed at the construction stage, price certainty is generally reviewed most carefully just before the decision to enter into a construction contract is taken. Whilst design, cost consultancy fees, and financing/client staffing costs may have incurred say 5 to 10 per cent of a construction contract value up to the date of receiving a traditional tender, the risk factor in the price certainty should then be confirmed or not at that stage.

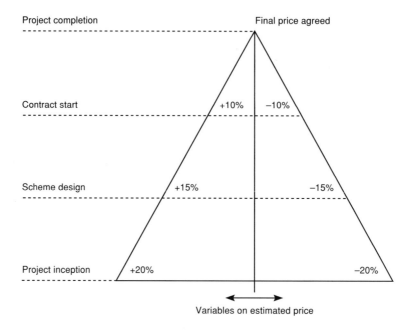

Figure 5.4   Change in price certainty during the life of a project

Different procurement routes will each produce different price certainty or uncertainty factors, at different stages of project commitment.

In *Thinking about Building*, price certainty is considered in answer to the question 'do you need to have a firm price for the project construction before you can commit it to proceed?' Categories of 'yes' and 'a target plus or minus will do' are given. It is emphasised again that price certainty is a relative, sliding-scale concept and with that in mind the following PAO will be appropriate, in the following order:

(i) Design and build
A design and build contract if correctly detailed and specified should establish a high level of price certainty. Changes due to design development should not normally be a risk to the client (this may be a factor present in traditional, separate design consultancy commissions). If a single-stage tender is appropriate for the appointment of a design and build contractor perhaps very few costs may be involved for the client, other than costs of preparing his 'employer's requirements', until a construction contract is signed. If a two-stage tender is appropriate, some design costs may need to be paid to a design and build contractor, in addition to costs of preparing the 'employer's requirements' and evaluating first-stage tenders, if the client does not enter into contract.

Variations of design and build such as develop and construct should provide a similar degree of price certainty, perhaps more so than that obtained from design and build tendering.

(ii) *Design and management*
Likewise this route should provide a high level of price certainty when operated say as a BOOT or PFI operation. If however client changes are introduced control of end price will become more difficult to achieve than with other routes. These systems cannot offer a high degree of price certainty unless the contractor offering design and manage has moved to become, at tender stage, the equivalent of a design and build contractor. Consultant design and management usually also cannot offer price certainty to the same degree.

(iii) *Traditional/sequential*
If design is completed by tender stage the construction contract should establish a high level of price certainty, leaving only perhaps fluctuations in labour and materials uncertain. This uncertainty should not be significant. Uncertainty over price does exist during the period from the feasibility/first-estimate stage until those essentially early-day budget costs have been confirmed by construction tenders. Costs necessarily incurred in consultancy fees are a risk in the traditional system. Price certainty is not always achieved in practice but that is generally due to entering into a construction contract when the design is not in reality 'firm'. Therefore significant variations are issued in order to achieve the required product and the contract price is often exceeded by the final account price.

(iv) *Traditional/accelerated*
The reason this route is chosen will principally be one of the overall programme time and design will not be so advanced under this route, therefore the degree of price certainty must be less. Even though approximate quantities, two-stage tendering and so on have become very commonly used the price certainty at contract stage must be less than obtainable with a full design contract. Prime cost contracts will be similarly even more uncertain at contract stage.

(v) *Management contracting*
It has been suggested (not least in *Thinking about Building*) that the management contracting system can offer price certainty. Management contracts do not offer price certainty because, at the time of contract, of necessity the exact nature and detail of the project are generally not established. Management is arguably a derivation of a form of prime cost contract and price certainty must be seen in this context. The device of a 'guaranteed maximum price' is sometimes offered, sometimes

used, but it is only possible to obtain price certainty if the maximum being guaranteed is high enough, in effect to contain a target figure that includes sufficient contingency. A maximum guaranteed price concept is not often possible to obtain before the time when a construction contract needs to be signed.

(vi) Construction management

This route will be, similarly to that of management contracting, unable to offer price certainty to a high degree because at contract commitment stage the total design is unknown. *Thinking about Building* agrees with this, putting price certainty into the category of 'a target cost plus or minus will do'.

Competition

Competition in building procurement is involved at every stage in the process. Building designers are now used to competing on quality of service, reputation and invariably fees. Building constructors have always been used to competing on price. The myriad of specialists and subcontractors ensure that competition is always available if desired.

Fair, comparable competition is an important constituent of competition in the UK. Codes of procedure for selection and appointment of contractors, using competition, are published by the National Joint Consultative Committee for Building and contain sound, practical advice on good practice. Competition on price is an important factor that should always be seen among the other criteria for competition.

Before a procurement route is chosen, discussions with the client should establish:

- how critical competition is to him
- what degree of competition for design, management and construction will be involved – competition for just one, two or all of these?
- how competition can be arranged consistent with other PAC.

Programme and possibly quality level and complexity of building involved may be the PAC that will restrict the amount of competition achievable.

A number of public clients will have standing orders, for reasons of satisfying public accountability, than often require them to obtain competitive tenders. Private clients mostly require competition for obvious commercial reasons.

Competition can be obtained using the following PAO:

(i) Traditional
In most circumstances, after the time and detail that will have gone into producing full design tender documents, selective competitive tenders are the logical outcome.

(ii) Design and build
Either one- or two-stage design and build will, if time allows, lead logically to selective competitive tenders.

(iii) Develop and construct
In develop and construct a consultant develops the scope design further, usually before competitive tenders are obtained.

(iv) Design and manage
This route will, similarly to the management route, demonstrate competition for the construction works.

(v) Management
Either management contracting or construction management involve and demonstrate price competition for appointing both the management organisation, as well as the works contractors that carry out the packages of work. This system clearly demonstrates, for the design team and client to see, competition for around 80 to 85 per cent of the construction contract cost.

If programme time or perhaps complexity are on balance more important PAC, the following PAO may still introduce some competition:

  (i) accelerated/traditional, perhaps with two-stage tendering and negotiation on the second stage
 (ii) two-stage design and build with negotiation on the second stage.

Management

In *Thinking about Building* (1985 edition) 'management' and 'accountability' were combined in essence into 'Responsibility', although the same intent was there, namely, can a client manage separate consultancies and contractors as opposed to managing one company? Management responsibility has become a most important PAC for many clients and is discussed further in Chapter 6. The rise in popularity of design and build is principally because management of design and construction is not divided in this procurement route after the briefing stage, and sometimes not before briefing either.

  A client may need fewer management resources if the number of companies working for him are fewer and, provided his consultancy and/or contract agreements are sound, responsibility concentrated in fewer places should be easier to manage and direct.

Accountability

If a client wants one organisation to be responsible and accountable for both design and construction then design and build procurement is the most appropriate. The limitations imposed on this route by other PAC have already been explored. Design and manage would similarly apply.

If a client wants to use his own team of consultants and contractors and can recognise, manage and accept the division of responsibilities so created the traditional and management routes will be appropriate.

How the procurement decision is made

As stated earlier in this chapter, it is sometimes useful, for other than the most simple procurement decisions, that a formal system of assessment is employed, perhaps using a numerical marking or assessment system. The BRT *Thinking about Building* does have a matrix system (reproduced as Figure 5.2 earlier in this chapter), without any numerical marking. Little other published information is available on a formalised approach to marking PAC.

Examples of a numerical marking system are contained in Building Procurement Systems, the Chartered Institute of Building, 1990, and also in an article called *'Which procurement system? Towards a universal procurement selection technique'*, Construction Management and Economics, E & F. Spon Ltd. R.M. Skitmore and D.E. Marsden, 1988. This last takes the NEDO PAC and applies weighted rankings to them. By statistical methods, it claims an advance first in accommodating disparate views of a panel assessing PAC and second in allowing for the interdependence of PAC.

A further example of a procurement assessment is that developed by the RICS in conjunction with the University of Salford in research funded by the Alvey Directorate. This work resulted in production of the 'Elsie' expert system ('Elsie' being derived from 'LC' meaning 'lead consultant'). One of the four modules in this system (currently being marketed under the name 'Imaginor') is a procurement module. The objective of this module is to advise rapidly on the suitability of different procurement routes and to identify a general procurement path that may best suit a client and his team. At present it is suitable only for considering most new building types and it is limited to traditional, two-stage traditional, management/construction management and contractor design and build.

It does not consider detailed contractual arrangements and looks at PAO according to potential for an early start of construction and the extent to which a contractor is to be involved in design and the extent to which a specialist management function is required. 'Elsie' is a welcome innovation and a worthwhile electronic aid in the area of procurement assessment.

The NEDO report *Faster Building for Commerce* contains an appendix (called an 'epilogue') entitled 'Dial your construction time'. This consists of two dials that attempt, very broadly, to allow calculation of construction time based on estimated value of the project, type of use of the building, whether it is to be purpose-built or speculative and whether it is a new construction or renovation. The NEDO 'dials' do not give guidance on a procurement route but they are of interest in helping with the assessment of construction periods.

The RICS publication *The Procurement Guide* gives advice, under section 4, on selecting a procurement strategy in relation to 'Time', 'Design/Performance', 'Cost' and 'Analysis'. These give a checklist by asking questions to establish the priorities of a client, along the lines of advice given in this chapter.

In practice a decision on which PAO to choose is often taken by a client, particularly if he is experienced in procurement, perhaps with advice from his principal source of external advice, sometimes at a quite early stage in a project. How the client came to be talking to his 'principal adviser', how the consultant or contractor became influential to the client in order to advise him, which discipline the adviser represents and so on is another matter. As stated in Chapter 2, the 1988 NEDO report *Faster Building for Commerce* recommends that advice from an 'independent customer representative' be made available from professional and trade bodies and this is an area where impartial advice from informed advisers is long overdue.

It is emphasised again that discussion of which PAO to choose needs to take place quite early in the life of a project because, as shown in Chapter 2, choices of consultants and/or contractors and the roles they each will need to play are affected by the PAO. The sequence of early decision is likely to be as follows:

- establishment of the market research and economic need to build. Very broad financial evaluations are needed to establish the worth or otherwise of proceeding with a project in more detail
- establishment of the likely quality, programme and price of the building product
- establishment of priorities and rationalisation in rank order of

- building product – technical complexity and quality of construction
- building programme – timing and flexibility
- building price – degree of price certainty and competition
- agreement of degree of complexity for assessing the PAC leading to a choice of PAO
- choice of PAO
- decision on consultant/contractor mix.

The case studies in Chapter 7 illustrate how decisions on procurement have been made in practice.

Summary

Before choosing a procurement route:

- all procurement assessment criteria (PAC) should be explored
- a priority listing should be made, principally to establish a balance between product quality, programme and price
- a formal assessment system may be appropriate
- it may be useful to place the procurement arrangement options (PAO) in rank order, for project record purposes
- organisations capable of carrying out the PAO that has been chosen should be known to be available to the client at the time that he requires them in order to satisfy the PAO chosen.

6   Contracts and Conditions

Introduction

This chapter deals with conditions of engagement for consultants, construction contracts, tendering requirements and procedures and related matters, referring where appropriate to other publications, standard forms of building contract and NJCC procedures for tendering, each of which cover more extensively the detail of the topics.

The influence of Latham

The Latham Report made fundamental recommendations on contracts for both consultants and contractors. In a comment on construction contracts Latham stated that 'endlessly refining existing conditions of contract will not solve adversarial problems'. He proposed instead that the contract drafting bodies, the Joint Contracts Tribunal (JCT) and the Conditions of Contract Standing Joint Committee (CCSJC) should coordinate their work and consider merger into a National Construction Contracts Committee, drafting any new contract around the NEC form. He recommended that effective reform of contracts should take into account a set of basic principles and be part of a family of interlocking documents (which the NEC has already attempted to do). The principles he recommended that a contract should contain included:

- a specific duty of parties to deal fairly
- firm duties of teamwork with shared financial motivation to pursue those objectives
- an integrated set of documents that defines the roles and duties of all involved, suitable for all types of project and for any procurement route
- contracts to be written in comprehensible language, with guidance notes
- separation of roles for contract administrator, designer, project manager and adjudicator
- choice of allocation of risks, decided as appropriate on each project and allocated to the party best able to manage, estimate and carry the risk

- taking steps to avoid changes to pre-planned work
- facility for pricing variations, when they do occur, in advance with provision for adjudication if agreement cannot be reached
- interim payments by payment schedules which carry through to subcontract payments
- short periods for interim payments with interest due for late payment
- secure trust funds for payment
- dispute resolution in first case by adjudication, litigation and arbitration only to be available after project completion
- mobilisation payments for off-site activities
- incentives for exceptional performance.

Latham suggested that the New Engineering Contract already contained most of these principles and he recommended that the industry move, over about four years, to use this contract for the majority of construction (that is, building and construction) projects.

Progress so far on the Latham recommendations for contract reappraisals has been slow. A few parts of the industry have taken to the proposed use of the NEC whilst other parts have decided to continue with the use and modification of more established contracts, both for construction and for professional services. The NEC has renamed its contract the NEC Engineering and Construction Contract (ECC) and with it is a contract for subcontracts (ECS), both reissued at the end of 1995. Some clients state their intention to use this and because of its 'new arrival' I have discussed its broad framework in more detail later. The Ministry of Defence is reported to be intending to adapt its existing contract forms rather than adopt use of the NEC. The Construction Industry Council (CIC) has produced its own pan-professional services contract. The NEC professional services contract has received mixed comments. Each professional institution currently publishes its own contract and recommends it for use by its members and their clients. The JCT and the CCSJC do not appear to be moving to form a joint body or to recast their contracts around the framework of the ECC.

The Construction Round Table has stated that its members (that include Amec Construction, BAA, Constructors' Liaison Group, Marks and Spencer, Alfred McAlpine, McDonald's Restaurants Limited) have the intention of committing themselves to 'fair construction contracts' and will give effect to principles that:

- deal fairly with those with whom they contract
- further working together and seek to adopt 'win–win' solutions to problems as they arise
- move towards the use of an integrated and compatible family of contracts which:

- are suitable for the full range of types of construction contracts and procurement routes
  - are written in easy to understand language
  - clearly define the roles and duties of everybody who contributes to the project
- allocate risks between the parties rationally so that each risk is taken on by the party that is best able to assess, carry and manage it
- minimise changes to the planned works
- base interim payments on agreed schedules of payments, milestones or activities
- make the period for interim payments as short as possible
- make late payment subject to interest
- provide rewards for exceptionally good performance
- minimise risk of dispute and provide for competent third-party resolution of disputes
- look to those with whom their clients contract to discuss any case where these principles have not been adopted
- commit all staff involved in construction to adopt these principles.

In essence these are very similar to the 'Latham principles'.

I have discussed the appropriate standard contracts in more detail below, in relation to each form of procurement, but because the ECC contract is new and lays claim to be appropriate for all methods of procurement, I have given general details of it here, as follows.

The NEC documents are published as a 'family' or 'suite' of forms within the following framework. The contract has:

- core clauses
- main option clauses
- secondary option clauses
- Schedules of Cost Components
- Contract Data Formats.

It is claimed that the ECC structure gives flexibility (the contract is held to be suitable for all types of procurement), clarity and simplicity (it is written in ordinary language – some lawyers have questioned this approach, suggesting that 'legal clarity' has not been achieved by writing in 'everyday clarity') and a stimulus to good management of a project.

The options for construction procurement are categorised as:

Option A – priced contract with activity schedule
Option B – priced contract with bills of quantities
Option C – target contract with activity schedule
Option D – target contract with bills of quantities

Option E – cost-reimbursable contract
Option F – management contract.

Parties to the contract are the 'employer' and the 'contractor' and roles within the contract are 'project manager', 'supervisor' and 'adjudicator'. These are provisions for 'subcontractors'.

Contracts

In order for a client or a principal adviser on procurement methods to make a decision, even a decision of the broadest principle, on a procurement path it is advisable for him to know the salient provisions of contracts for professional services as well as those of contracts for construction. It is interesting to note that *the client* when contracting with a builder generally becomes *the employer* and *enters* into a *building contract*. When *the client* contracts with a consultant he quite often remains *the client* and enters into a contract, called for instance *an appointment, an agreement* or *conditions of engagement*. In principle there is no difference – 'contracts', 'appointments', 'agreements' and 'conditions' are all contracts. Like perhaps marriage, 'conditions of engagement' should be seen as a contract and should not be seen as rules for engagement, that is, as a term used for adversarial actions, in for instance military or naval engagements.

Although contracts for building and contracts for professional services are different in purpose they should essentially be complementary. The fact that they are not always so is part of the historical lack of coordination and consistency of contracts. Both within and between the various professional services contracts, and in turn between those professional services contracts and the range of building contracts there is lack of consistency and coordination. The wording of a building contract can condition the status and authority of a professional acting within that building contract, although the professional is not employed by the building contractor, and the client's contract(s) with his professional(s) should recognise this. The actions required of a professional by a building client, for instance during the pre-construction stages of a project, may also be influenced by the procurement route to be adopted and therefore by the responsibilities taken on by the contractor. Naturally contracts for professional services provided to a construction contractor may be somewhat different again.

Contracts for professional services

Contracts for design consultancy and cost consultancy services are most important for building clients and for this reason more attention is now being paid to them. As market competition for consultancy services has increased, including competition on fees, the services offered and the contract conditions for provision of those services have both come under review. Whereas at one time it was commonplace for a professional institution/institute's conditions of engagement, together with their scale of fees, to be offered and accepted, both of these have now often become matters for discussion, for negotiation, amendment and competition.

Architects, engineers, quantity surveyors and project managers are the principal professionals that will be involved in providing professional advice in the process of building design and construction, including perhaps advice on building procurement. It is becoming quite common for some of the larger clients to have their own conditions of engagement, that is, a contract, for appointing building professionals. It is also now not uncommon for major consultancies to have their own conditions.

A draft of a consultant's agreement, common to all building consultants, has been produced by the Joint Contracts Tribunal. At the date of this edition, consultation on this is beginning throughout industry bodies.

Professional design and cost consultancy contracts

If a contract for professional services is not drawn up especially between a client and a consultant the following 'standard' conditions are the usual contracts used for the employment of the principal professionals by a client:

- architect – *Standard Form of Agreement for the Appointment of an Architect*, published by the Royal Institution of British Architects (RIBA)
- consulting engineer – *ACE Conditions of Engagement*, published by the Association of Consulting Engineers (ACE)
- quantity surveyor – *Form of Agreement and Standard Conditions of engagement for the appointment of a quantity surveyor*, published by the Royal Institution of Chartered Surveyors (RICS)
- project manager – *Memorandum of Agreement between Client and Project Manager and Conditions of Engagement*, issued by the RICS and the Project Management Diploma Holders Association of the RICS. Alternatively the *International General Rules and Model Form*

of Agreement for Project Management, produced and issued by the International Federation of Consulting Engineers (FIDIC), is often used as a basis for project management consultancy

- pan-professional services – issued as part of the NEC family of documents for use by any professional
- professional services contract – prepared by the Construction Industry Council. This contract, designed to be used by any building consultant, was prepared by the 'Harmonisation of Conditions of Engagement Steering Group' of CIC and at the time of preparing this edition it is understood may be published in 1997.

The above contracts detail services for design, cost consultancy and project management and they contain a large range of services.

Procurement advice

The advice that a client *needs* on procurement arrangement options (PAO) will be governed by the following:

(i) the client's knowledge of the construction industry
(ii) the expertise of any in-house project executive.

The advice that a client *receives* on PAO may depend on the following:

(i) points (i) and (ii) above
(ii) the involvement of any external advisers to the client
(iii) the timing and the terms required for appointments.

Procurement advice can be given both by consultants and by contractors and its impartiality should always be looked at. Ideally the advice given by any organisation on PAO should be impartial so as 'to ensure that advice is not simply intended to maximise workload of the adviser's organisation' (as stated in the NEDO version of *Thinking about Building*) – it might also be added, or to minimise it.

Standard building contracts with contractors do not contain, nor would it be appropriate for them to contain, provisions for obtaining procurement advice. Standard contracts cover construction activities and on occasions design and management activities. Contractors may and can provide advice on procurement but it is important to bear in mind the extract on impartiality, from *Thinking about Building*, quoted in the last paragraph.

Standard contracts for architects, consulting engineers and quantity surveyors may not refer directly to procurement but they may do so indirectly, in the sense of 'reviewing' procurement. Consultants may each offer advice in the course of reviewing design and construction arrangements with the client but they are not, reading the fee scales and services agreements together, solely responsible to the client for the procurement decision.

In a more active, leading sense, contracts for architects, consulting engineers, surveyors and project managers may include a direct responsibility for procurement advice, for instance, as follows:

- the RIBA *Conditions of Engagement for the Appointment of an Architect (CE/95)*, under 'Schedule of Services to be provided by the Architect', inception and feasibility, 'advise the client on methods of procuring construction'
- the *International Federation of Consulting Engineers (FIDIC) Model Form of Agreement for Project Management* has provisions to enable the supply of project management services of 'the management of planning, design, procurement, construction, management and commissioning of a capital project'
- the RICS *Project Management Agreement Conditions of Engagement*. Guidance Note 13 refers to 'Contract Procedures' and Note 13.2 states, 'Decide with Consultants procurement procedure for appointment of contractors. Decide on type and form of contract. Monitor Consultants in the preparation and assembly of tender documents. With the Consultants, check the form and content'.

The cost and value of procurement advice

From the above provisions it can be seen that the usual contracts for the provision of design services, cost of consultancy services and building construction do not include project management services nor do they provide inherently for direct procurement advice, selection of a procurement route or implementation of that route. They often do include reviewing procurement arrangement options, generally in concert with the client and other consultants. If specific, impartial advice on procurement is wanted for other than simple projects, involving simple choices, then the contract to provide such advice should be specific. Legal advice may be required if standard arrangements either are not available, or are not appropriate or have to be amended significantly.

Project management services are usually specific to each project and justify an agreement to reflect that. The responsibility for project management can be quite onerous and it is for this reason that many organisations offer project coordination instead.

Professionals should not readily take on responsibility for the selection and implementation of design and construction procurement arrangements without a clear authority, and without a contract and suitable fees.

Contracts for building

Contracts in general contain provisions and allocate a balance of responsibility, reward and risk. Construction contracts have been produced to suit the procurement methods required and in so doing they place the balance of responsibility, reward and risk accordingly on the parties to the contract. Each contract may amend that balance and construction contracts can be classified according to two fundamentals, namely according to the allocation of responsibility and also according to the basis of reward. Risk can then be judged from where responsibility is placed and also from how reward is agreed to be paid.

Responsibility

Construction contracts can be classified by how design and how coordination, or management, is placed.

A decision must be taken, usually quite early in the review of PAC, on:

- design responsibility – design (see glossary) is an inescapable part of construction procurement and a decision must be taken on how, if at all, design responsibility is to be apportioned over more than one organisation

and

- coordination responsibility – construction (see glossary) is another inescapable part of construction procurement and a decision must be taken on how, if at all, responsibility is to be placed for coordination (see glossary) of the process of design and construction on to one organisation or between a number of organisations.

Reward

Construction contracts can also be classified by how reward (payment) is paid to the contractor by the client.

A decision must be taken, usually fairly early in the review of PAC, on:

- a price-based contract – these contracts have payment based on the price quoted by a contractor for his work. The price quoted may or may not relate to the cost of that work. Additional work ordered within a contract is generally evaluated and paid for in relation to the basis of the contract price or

- a cost-based contract – these contracts have payment based on the actual cost of construction. However much work is carried out is paid for on a pre-arranged basis of reimbursement of costs, plus probably arrangements for payment of overheads and profit.

Risk

Responsibility and reward appear to be two parts of a simple equation. Risk is the problematic, missing part that arises from judging responsibility and attempting to price it in order to obtain reward.

There will always be some risks that cannot be quantified and cannot sensibly be priced and risk is generally covered in construction contracts as follows.

Fundamental risk

Fundamental risk is risk such as the occurrence of war, nuclear happenings, aircraft happenings and similar events. These risks are all covered by government statutes and construction contracts usually refer to them, and place the effects of them if they did occur, outside of the contract. Construction contracts cannot price for the fundamental risk of such a 'happening' and no insurance is normally required within the contract for such risks, or in fact could normally be obtained.

Pure and particular risk

Pure and particular risk is risk such as injury to persons and damage to a building or to neighbouring buildings during construction due to fire, storm, collapse, subsidence, vibration, water escape and similar events. Construction contracts usually require the contractor to indemnify the employer against some of the effects of such happenings and they also generally require the contractor to take out insurance to cover some of the effects.

Speculative risk

Speculative risk is risk that can be varied in its incidence between the parties as they wish.

For instance the following may be placed in a contract as being outside the control of either party:

- *force majeure*
- exceptionally bad weather
- damage by fire and similar perils (although insurance to cover

rebuilding may be required, the effects and responsibility for the damage will be outside the contract)
- industrial action
- non-availability of materials and labour
- delay by specialist contractors and statutory authorities.

The following may be placed in a contract to be within the control of the employer:

- variations
- delay in giving instructions to the contractor
- discrepancies found in contract documents
- postponement of site activities
- any result of work carried out on the site by the client's other contractors
- work resulting from provisional sums
- interference with the contractor's method of working and/or use of the site.

All other matters not defined above may be placed to be within the control of the contractor.

The losses involved due to the above events may be in time and/or in money. Although some of the above-mentioned events may in fact be beyond the control of the contractor he may nevertheless be asked to bear any losses, both to himself and to his client, because of those events or he may be asked to bear only his own losses or no losses at all. The detail and balance must be exactly described and apportioned in the contract documents. In this way, difficult though it may be, the contractor and the client can attempt to assess the probability of an event occurring and then to price (if it is the contractor) or to allow (if it is the client) for its effect.

The speculative risk of the client and contractor varies according to the type of procurement adopted and the type of contract chosen. Figure 6.1 shows the range of risk.

It can be seen from the earlier sections on 'Responsibility' and 'Reward', that a number of contract arrangements are possible and that the allocation of risk arises from the contract chosen. Clearly it is probable that more risk lies with a contractor if he has a contract for design and build than if he has a contract where his costs are reimbursed. If a contractor has a price-based contract and has quoted before work is constructed he will wish to see whether or not the work is the same work for which he tendered and whether or not it is carried out under the same conditions as those he understood would be applicable when he tendered. If a contractor has a cost-based contract, including an element

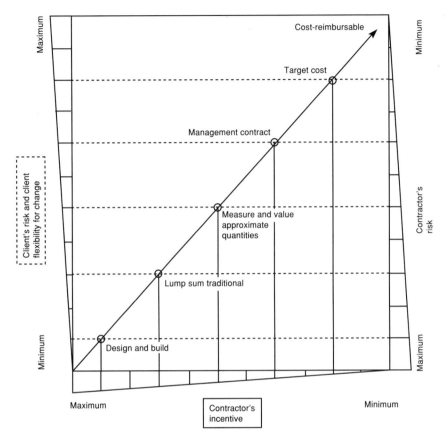

Figure 6.1   Procurement risks and characteristics of types of building contracts

for overheads and profit, he may have little concern for how, where or when the work is carried out but the client may be most concerned because the price risk has been transferred to him in the contract.

Before examining construction contracts in more detail it is emphasised that a PAC review is essential before leading to a choice from the PAO. A procurement choice will then lead to the type of contract and then to a particular form of contract.

Price-based contracts

Within the price-based group of contracts there are:

- *Lump sum contracts*

A lump sum contract is when the contract sum is agreed before construction starts and the price is stated in the contract, often as 'the contract sum'.

Lump sum contracts are appropriate for traditional (construction only) contracts, design and build contracts (sometimes then called 'turnkey' or 'package deal' contracts), and works contracts (within management contracting or construction management contracts).

The information supplied to the contractor can vary along a range from, at one end, a performance statement up to, at the other end, very detailed specifications of materials, workmanship and drawings, often additionally with a bill of quantities.

A presumption in the drafting and in the use of a lump sum contract is that there is reasonable expectation in the minds of the contracting parties at contract stage that there will be no substantial, fundamental variations issued and that certainly circumstances, as known at the outset of the works, do not lead to a conclusion that remeasurement of the completed work is most likely.

Within lump sum contracts there are generally provisions for issuing variations and for valuing them according to valuation rules that relate to an agreed price basis. Lump sum contracts can transfer the maximum amount of responsibility and risk to a contractor. In some cases this can or should restrict the client's freedom to make changes as he may then be in a relatively weak negotiating position.

Lump sum contracts provide to a client the maximum price certainty before construction commences, provided a client's requirements have been fully specified in the contract.

• Measurement contracts

A measurement contract is when the contract sum is only established with certainty on completion of construction, when remeasurement of the quantities of work actually carried out takes place and is then valued on an agreed basis.

Measurement contracts are sometimes referred to as 'remeasurement contracts'. They are used when the work required cannot be accurately measured for the tender bill of quantities. The most effective use of a measurement contract is where the work has been substantially designed but final detail has not been completed. Here a tender based on drawings and bills of approximate quantities will be satisfactory. Sometimes measurement against a schedule of rates (that is, prices) for categories of work is appropriate. Measurement contracts allow a client to shorten the overall programme for design, tendering and construction but usually with the result of some lack of price certainty at contract stage because the contract quantities reflect the lack of information at tender stage on exactly what is to be built. The scope of the work, the approximate price and a programme should be clear at contract stage. Measurement contracts provide more risk than lump sum contracts for a client but probably with programme advantages.

Cost-based contracts

Within the cost-based group of contracts there are:

* *Cost reimbursement contracts*

Cost reimbursement contracts are used when the contract sum is arrived at on the basis of actual costs of labour, plant and materials used in the works plus an agreed allowance for overheads and profit.

Cost reimbursement contracts are used when it is not appropriate to measure even approximate quantities because either the scope and nature of the work is not clear before construction must start or the risk inherent in the project is such that it is unreasonable to ask a contractor to price it (and it follows that no contractor will wish to price the work).

Examples of this would be emergency building repairs following a fire, some elements of renovation and some civil engineering work, for instance where relatively small amounts of work are needed under water, or where subsidence has occurred and needs urgent attention. Cost reimbursement contracts provide for great flexibility for change in programme, scope and quantity of work. The risk to the contractor is low, except perhaps in relation to his fee, but the risk to the client is high.

Because of the weaknesses of cost reimbursement contracts in providing an incentive to the contractor to try to minimise his 'prime costs' for the construction works there are several variants on the system, each one depending on the way the fee (for management overheads and profit) is applied.

* *Cost plus percentage fee*

The prime cost is reimbursed as described under cost reimbursement above plus a flat, or reducing, sliding-scale percentage fee. This is not much of an incentive to the contractor to contain his prime costs.

* *Cost plus fixed fee*

The prime cost is paid plus a fixed fee. This may be more of an incentive for the contractor to work efficiently.

* *Cost plus fluctuating fee*

The fee paid to the contractor fluctuates as a proportion of the difference between the estimated prime cost, calculated before any work is started, and the actual prime cost of the construction. The fee decreases as the cost increases. This, it is argued, may restrain any contractor's supposed inefficiency.

• *Target cost reimbursement*

The fee is based on an agreed target estimated for the prime cost of the work and the relationship of the actual with the estimated prime cost affects the fee. Setting the target correctly is the difficult part. If the target is set too high the incentive for efficiency is reduced, if it is set too low the fee is adversely affected. In theory the system is attractive but in practice it can be difficult to achieve the objective of an incentive but with a large prime cost element.

Types of contract

The types of contract, namely price-based and cost-based contracts, that have been previously described are sometimes applicable to a number of procurement methods, but in some cases only to one method. It is therefore appropriate to review the types of contract within the categories of procurement and to look at them under the PAC described in Chapter 5.

Contracts where design is combined with construction – design and build procurement

For this procurement method, which includes the variation of develop and construct, the considerations are:

Programme (timing)

Programme should be relatively guaranteed. One of the reasons to use design and build is programme. It can be a faster method than the traditional route and programme should be a speculative risk that the contractor should take as, in principle, he is in control of design as well as of construction.

Controllable variation

Variations should not occur if the employer's requirements are comprehensive. However, if variations are introduced there may be heavy price penalties.

Complexity

A contractor should have assessed the nature of the building that he is to provide. If a client requires a complex building he should have taken this into account before deciding on design and build procurement and therefore which contractors to invite to tender.

Quality level

A client usually has no direct control over quality once the contract standards of quality have been accepted and he must find out at tender stage how the contractor's proposals do or do not meet his requirements. Site supervision/inspection by a client's staff can be used to monitor a contractor's obligations on product quality.

Price certainty

Price should be relatively guaranteed. There is usually no provision for a bill of quantities and the contractor bears responsibility for the accuracy of his tender. Because there are no bills, adequate arrangements for evaluating any changes on a price or cost basis must be included in the contract.

Competition

It is often difficult to evaluate competitiveness of design and build because different proposals may need to be compared. Any advantages of competition need not necessarily all be passed on to the client but it must be assumed that in principle they are.

Management

One of the attractions to the client of design and build contracts is placing the responsibility for both design and construction on to the contractor. The balance of how much a 'scope design' may have prescribed the design may limit that overall responsibility.

A client retains a responsibility during the contract through his 'employer's representative'.

Risk avoidance

A contractor usually takes on many of the speculative risks, as defined in this chapter. The consequences to a client of negligent design should be considered and possibly covered by insurance, especially if design consultants have been employed by a design and build contractor, and if the level of design liability has been restricted by the contract. Warranties are discussed later in this chapter.

Forms of contract

A design and build or develop and construct contract is by its nature a 'lump sum contract'. The standard forms of contract suitable for design and build and develop and construct are:

- *Joint Contracts Tribunal, Standard Form of Building Contract with Contractor's Design*, 1981 Edition. This is a form capable of obtaining either design from a functional or performance brief, or development of design and then construction when a 'scope design' is provided to the contractor by a client
- Joint Contracts Tribunals. *Standard Form of Building Contract with Quantities*, 1980 Edition, *Contractor's Designed Portion Supplement* 1981. This form is suitable if selected elements only of a building are required to be designed by a contractor, rather than all the elements of a building when the *JCT with Contractor's Design contract* should be used. Consideration is now being given to incorporating provisions currently in the design portion supplement into the *Standard Form of Building Contract*, so making a supplement unnecessary
- Forms of contract adapted for use with the British Property Federation System for building design and construction
- NEC Engineering and Construction Contract (ECC) 1995, option A – priced contract with activity schedule
- GC/Works/1 (Edition 3) single-stage design and build version (1993). This was developed for use by public service organisations but could be used for private procurement.

Contracts where design is separate from construction – traditional ('lump sum') procurement

Programme (timing)

Programme is not likely to be either the shortest or the most certain part of this procurement method. The overall period for sequential design followed by construction is likely in the majority of cases to result in the longest procurement route. Accelerated, two-stage or negotiated tendering may shorten the overall programme time, but probably with some associated risks. Provided that all design information were to be completed before tendering, a construction programme should then become fairly certain in a contract.

Controllable variation

A client and his consultants control the origin of variations and they should not occur if pre-construction design has been good. If variations have to be issued traditional contracts provide a comprehensive framework for valuing them according to well-known rules.

Complexity

Simple or complex buildings can be provided with traditional contracts. Provisions for selecting specialist subcontractors are usually included.

Quality level

A client has control of quality by the standards that he specifies and also through inspection by his consultants and perhaps by site supervisory/inspection staff, appointed to see that he obtains those standards.

Price certainty

Provided the tender documents are complete price should be relatively certain before construction starts. The price of variations is agreed according to well-known formulae.

Competition

The contract sum is generally the result of competition.

Management

Traditional contracts are generally clear that the contractor is only responsible for construction. Design, other than by consultants, does sometimes become part of the construction contract through specialist design. Then responsibility for design is less clear and has led to the growth of client/contractor or specialist design warranties.

Risk avoidance

Traditional contracts generally place the balance of risks fairly evenly between the parties.

Forms of contract

The standard forms of contract for traditional procurement are:

for lump sum contracts
- Joint Contracts Tribunal, *Standard Form of Building Contract*, 1980 Edition
- Joint Contracts Tribunal, *Intermediate Form of Building Contract for Works of Simple Content*, 1984 Edition
- Joint Contracts Tribunal, *Agreement for Minor Building Works*, 1980 Edition
- *NEC Engineering and Construction Contract* (ECC) 1995, option A – priced contract with activity schedule – or option B – priced contract with bill of quantities
- *GC/Works/1* (Edition 3) lump sum with quantities (1989) revised 1990 and lump sum without quantities (1991). This was developed for use by public service organisations but could be used for private procurement
- *PSA/1* (1994) contract.

for measurement contracts
- Joint Contracts Tribunal, *Standard Form of Building Contract*, 1980 Edition with Approximate Quantities
- Institution of Civil Engineers, *Conditions of Contract and Forms of Tender, Agreement and Bond* for use in connection with Works of Civil Engineering Construction ('the ICE Conditions of Contract') Sixth Edition, 1991
- Institution of Civil Engineers, *Conditions of Contract for Minor Works*, First Edition, 1988
- *NEC Engineering and Construction Contract* (ECC) 1995, D – priced contract with bill of quantities
- *GC/Works/1* (Edition 3) lump sum with quantities (1989) revised 1990 and lump sum without quantities (1991). This was developed for use by public service organisations but could be used for private procurement.

for cost reimbursement contracts
- Joint Contracts Tribunal, *Standard Form of Prime Cost Contract*, 1992
- *NEC Engineering and Construction Contract* (ECC) 1995, option E
- *GC/Works/1* (Edition 3) prime cost form of contract (1990). This was developed for use by public service organisations but could be used for private procurement.

Contracts where design is separate from construction – fee management procurement

Programme (timing)

Programme, both for overall completion and for an early start of construction, is a prime reason for using this form of contract. An early start of construction can be made by letting subcontracts to works contractors, often before some of the later work has been designed.

Controllable variation

These contracts generally expect variation to be necessary and that lack of pre-design will be a constituent part of the work. Design therefore sometimes has to be modified or developed during construction and a management contractor will adjust his programme and price or cost accordingly.

Complexity

Management contracts were developed to suit complex design and construction and require appropriate management skills to achieve success.

Quality level

As with traditional contracting a contractor is usually responsible for meeting specified standards of quality for materials and workmanship. Client's site inspection/supervisory staff may also need to be considered.

Price certainty

These contracts generally provide for a start of construction based on a cost plan, preliminary drawings and specification. Price is only known with certainty at completion and initial indications of price have to be viewed as just that. Because of the use of many subcontractors (works contractors) the interactions of their activities one with another can produce variations arising from programme changes.

Competition

Competition is often part of the selection process to appoint a management contractor and is usually one of the main selection criteria to appoint works contractors that actually carry out construction and install the environmental services.

Management

Responsibilities for completion to programme, price and product quality are spread over a number of organisations and success depends very much on the management contractor's or construction manager's skills.

Risk avoidance

Management contracts or construction management arrangements leave the majority of speculative risk with a client. They rely on the client's ability and willingness to have a management organisation expend resources on behalf of the client to obtain programme completion, often at the expense of price certainty and/or possibly quality of finished product.

Forms of contract

The standard forms of contract for management procurement are:

for management contracts

- Joint Contracts Tribunal, *Standard Form of Management Contract,* 1987 Edition
- Joint Contracts Tribunal, *Standard Works Contract* 1987, between a Management Contractor and his various Works Contractors.
- *NEC Engineering and Construction Contract* (ECC) 1995, option F.

for construction management

- There is at present no JCT standard form of contract for the provision of construction management although one is in draft. Several companies that offer this procurement form have their company standard and a client should obtain advice from his principal adviser and possibly also from a lawyer experienced in construction contracts.
- *NEC Engineering and Construction Contract* (ECC) 1995, option F.

Collateral warranties

The concept of a collateral warranty is simple. It is an agreement that runs parallel or collateral to another agreement. It can run alongside a contract for building or alongside a contract for professional services, these being the contracts between a client on the one hand and a contractor or a consultant on the other hand. A warranty is an agreement between a contractor or a consultant on the one hand and a third party (who is not a party to the contract between a client and a contractor or between a client and a consultant) on the other hand. A client is not a party to the warranty. It is common for a third party to be a financier or a prospective purchaser of a project or a tenant of the finished building or parts of that building.

The purpose of a warranty is to give a third party a direct, contractual right of action against a contractor or against a consultant in the event of a building being defective in any way.

The reason for the emergence of collateral warranties is because the present state of the law has made recovery of loss through the route of tort less satisfactory and easy to achieve than some have wanted. Some clients, principally property development companies, have been pressed and have therefore sought to arrange direct contractual relationships for their financiers, purchasers and tenants by providing collateral warranties with the contractors and consultants that provide the building design and construction to a client.

The different positions of third parties, contractors and consultants that currently exist are described shortly. It is emphasised that events and case law, and their effect on tort, on contract and in particular on the need for detailed provisions contained in collateral warranties may change in this area and legal advice should always be sought as appropriate. The relationships of client, purchasers, tenants, contractors and consultants are shown in Figures 6.2 and 6.3.

A *Form of Agreement for Collateral Warranty* was prepared in 1992 and approved for use by the British Property Federation, the Association of Consulting Engineers, the Royal Incorporation of Architects in Scotland, the Royal Institute of British Architects and the Royal Institution of Chartered Surveyors after discussion with the Association of British Insurers. It is for use where a warranty is to be given to a company providing finance for a proposed development and to a purchaser or tenant of premises in a commercial and/or industrial development.

The third party's position

Because of the limitations inherent in the scope of a duty of care owed by a consultant, under the law of tort, third parties usually seek to improve their position through the law of contract. The need for such a contractual relationship may be particularly relevant with the requirement of current commercial leases that often impose the entire responsibility for repairs on to a tenant, thereby excluding any remedy against the landlord in the event of defects in the design or construction of a

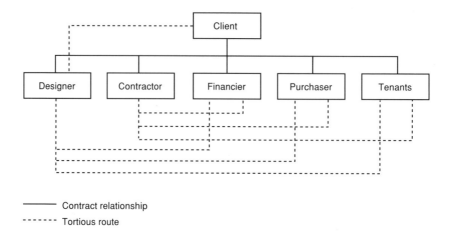

Contract relationship
- - - - - - Tortious route

*Figure 6.2   General contractual and tortious relationship
of organisations*

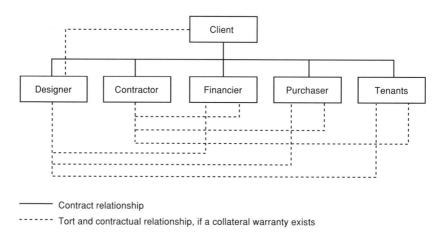

Contract relationship
- - - - - - Tort and contractual relationship, if a collateral warranty exists

*Figure 6.3   General contractual relationships of organisations,
including those established by collateral warranties*

building. A tenant or financier may be left to bear the whole cost of
remedying defects of workmanship or of design if he did not have any
direct right of action against a contractor or consultant, that is unless
he could obtain redress through the law of tort for negligence. Action
through tort has recently been found less easy and certain than action
through a contract, namely a warranty. A further reason for seeking a

warranty is that a contractor may also become directly responsible for his subcontractors under a collateral warranty.

The contractor's and consultant's position

Under standard forms of construction contract a contractor is generally under no obligation to enter into collateral warranties but such a requirement may be made as an amendment to a contract. Standard contracts often allow for assignment of contractual rights but only with the consent of the other party, not to be unreasonably witheld. In any case it is essential, commercial sense that a contractor does not undertake obligations under a warranty that are beyond and more onerous than those in his contract with the client.

Consultants may be in a similar position to that of a contractor. No institute or institution's conditions presently incorporate warranty provisions and before entering into a warranty a consultant should consider whether or not the obligations proposed in a warranty go beyond the duties and responsibilities that he had undertaken in his contract of engagement with his client. Most importantly, he should also ascertain that any warranty he enters into will be covered by his professional indemnity insurance.

Tendering

Tendering is the name given to the process or procedure that is used to obtain offers leading to a contract between a client and contractor, a client and consultant, a contractor and subcontractor, and so on. In order to start a tendering process a decision will need to have been made, positively or otherwise, on the contract arrangements that will have to be entered into on acceptance of a tender. Thus 'tendering' and a 'contract' are clearly distinguishable in concept but are very closely linked in practice. Some people will talk about a 'negotiated contract' when strictly they mean a 'negotiated tender'. Some people talk of an 'extension contract' when strictly they mean a new, separate contract, arrived at by competition or by negotiation from the basis of an existing contract or contracts.

Open tendering

Open tendering is a procedure of allowing virtually any organisation, probably following advertisement(s) through technical, trade and/or the general press, to submit a tender. Advertising will give outline details of the work, its scale, programme, probably the form of contract and other relevant matters. At its simplest the apparent appeal of open tendering is that it should secure maximum competition by giving any organisation an opportunity to tender. It is strongly discouraged by bodies such as the NJCC and its use has long been in decline. It is not considered good practice for construction work as it can lead to organisations tendering that do not have adequate resources and/or expertise.

Selective tendering

Selective tendering is a process or procedure for selecting a limited number of firms and then inviting them to tender.

The NJCC has published three codes on selective tendering. These are for single-stage and two-stage tendering, appropriate when using complete design information, and for design and build tendering (which includes single or more stages of tendering). A code for selection of a management contractor and works contractors was published in 1991. In addition NJCC Guidance Note 1 details requirements that need consideration when joint venture tendering is appropriate.

Selective tendering needs careful application but it is the recommended method that in most cases is appropriate where competition is wanted.

Single tendering

In some cases only one organisation can satisfy the criteria for selection to tender. For instance a specialist subcontractor may have a unique product. Similarly the design skills of a contractor, in the circumstances of a tender, may make his offer unique. In these cases facts should be faced and it may be more appropriate to negotiate a tender rather than to receive a single, non-competitive tender.

Negotiated tender

Usually in tendering one party makes an offer to the other party as a result of its assessment of technical and commercial aspects, based on the information provided as a basis for tendering. In some procedures for tendering the offer is then generally accepted, or rejected, or alternatively negotiation may follow in some cases. Incidentally, it is good practice that the tendering conditions have indicated how and when negotiation is possible.

Generally in negotiation the two parties go through a procedure of proposal and counter-proposal. This may involve progressively moving forward in separate areas, perhaps reassessing periodically each aspect of the negotiation and then bringing all the parts together, in order to make an interim assessment of progress on how the whole negotiation is proceeding. Neither party will be committed finally until the pattern of the whole tender is reached, when it is then open for acceptance or rejection in its totality.

Procedures for negotiation, endorsed by consultative bodies such as the NJCC, after single-stage competitive tendering, envisage negotiation with one firm only after the stage of assessment of the competitive tender. If a process of two-stage tendering is to be adopted, the procedures for negotiation should be made quite clear in the tender inquiry. In some cases first-stage tenders can be virtually statements of intent, and negotiation from the intent to the detail may then be carried out with one or more organisations. There may be other circumstances when parallel negotiation with several companies, or successive negotiations with one or more organisations, are required if discussions with one firm run into problems. Temptation to develop 'a dutch auction' should always be resisted.

As the range of procurement options has increased beyond the traditional path of completing all design and then tendering for construction, probably with bills of quantities, the requirement for the use of negotiation has also increased. Negotiation may now arise from the evaluation process of assessing incomparable tenders for:

- design and build
- two-stage tendering for construction only, with more than one tenderer proceeding to a second, or more stage(s).

Negotiation is a skilled process and requires skilled clients, consultant advisers and contractors and those participating in it should be adequately trained and experienced.

Extension contracts

Contracts that arise from the circumstances of, and often using the arrangements of, a previous contract are often referred to, loosely, as 'extended' or 'extension contracts'. Whereas a contract can of course by agreement literally be extended, that is, the original contract is increased in its scope and/or intent by a variation to it, the colloquial use of 'extension' is often as follows. The phrase is generally used when there is a desire to place another, new contract with an organisation that is already engaged in an existing contract, either on the same site or perhaps carrying out work of a very similar nature elsewhere. An 'extension contract' may be made competitive or more usually it is agreed following negotiation. Benefits of continuity and/or repetition of work and thereby increased speed of working are claimed for both client and contractor.

Serial tendering

'Serial tendering' is the term used where tenders are invited for a project or series of projects, each of which will become a contract or contracts in their own right. It is a predetermined form of 'extension contract'. The tendering instructions and/or contract arrangements for the first contract generally will contain provisions to enable the formation and letting of the contracts that are intended to follow on.

The clear intention of a serial arrangement should be that both parties carry through the whole series of projects on the basis laid down at the contract stage of the first contract. There should be clear provisions set down to allow for unexpected developments that may occur when the intention to proceed into the second and/or subsequent projects either cannot or, for equitable reasons, should not carry on.

Summary

A client may need to enter into contracts for:

- professional services to provide procurement advice, project management, design and cost consultancy
- construction services to erect a building, including or excluding its design.

Contracts for professional services and construction services will vary according to the procurement option chosen, the salient points being:

- where responsibility for design is to be placed
- where responsibility for coordination of the design and construction process is to be placed
- how reward is to be assessed, based either principally on a cost basis or on a price basis
- how risk is to be apportioned between a client, consultants and contractors
- standard contracts exist for the main procurement options
- collateral warranties or third-party warranties are appropriate in some cases in order to allow a client or a third party to have a contract with a consultant or contractor.

Tendering, according to well-known procedures and good practice, should enable selection and appointment of consultants and contractors to be made appropriate to each of the procurement arrangement options.

Part 2

7 Case Studies

Introduction

The case studies in this chapter illustrate the principles of procurement assessment criteria (PAC) and procurement arrangement options (PAO) that have been discussed in earlier chapters.

Studies have had to be hypothetical cases and, in some respects therefore, artificial. It is attractive to think that case studies could be followed exactly as they occurred in reality, but there are difficulties with this approach. First, if named projects, named clients, consultants and contractors were to be used, permissions of all the principal parties involved would probably be required. Second, the views of each party involved would probably need to be given, with qualifications from a number of parties perhaps, if any explanation or possible defence of a particular point of view or party seemed necessary.

Clients may often be sensitive about the role that they play and usually, quite understandably, want exposure in at least a favourable, if not the best possible, light. Consultants and contractors may also feel keen to keep their image as they perceive it and/or want it to be seen, that is, naturally, in the best possible light. Consultants and contractors may feel this even more strongly than do their clients and they do not readily wish to have any possible suggestion made that their advice and their actions were ever anything other than detached perfection, given at the right time without fear or favour. The case studies are therefore an amalgam of actual projects, chosen to illustrate the challenges which procurement decisions must resolve. Some of the realistic details of actual projects have been omitted and other fictitious parts have been introduced, to illustrate problematic areas.

A consistent format for each of the fourteen case studies has been adopted in order to reinforce the commonality of establishing PAC that must first be drawn out before PAO can be deduced. Details of the client, the project, the main (initial) considerations, the PAC and PAO, the reasons for the procurement choice and a postscript are given.

Clients and practitioners, experienced in procurement advice and decisions, will be aware that the logical approach recorded in these case studies does not in practice always proceed along a sequential path. Students, inexperienced in construction procurement, will hopefully find the case studies easy to follow because of the commonality in their presentations. However they should always remember that, in condensing the PAC and PAO for case presentation purposes, many tangential, non-sequential issues that occur in practice may have had to be resolved in order to keep the process of procurement decision moving along its path. In order to keep the case studies within a reasonable length they may have, on occasion, suffered from excluding more of the many peripheral, human issues that in real life, with its complex interactions, often intrude. The common approach has been repeated in each case study and so allows each study to stand alone and to be used as an exercise complete in itself, without the need for reference to other case studies to illustrate comparisons. For this reason reading all studies straight off should enable a student, on occasions, to reach the conclusion before he reads it.

The choice of a procurement path is one thing, the adherence to that path as the project progresses is another. It is most important to explore PAC sufficiently long and hard in order to establish PAO and then to make the correct procurement choice. In practice the problem still remains that clients with their advisers, consultants and contractors decided their priorities but then sometimes they change them. Generally it has to be believed that procurement choices are made in all good faith after balancing the variants of price, programme and product. If a procurement choice has been made in good faith, perhaps like the modern statements of good faith in oral marriage contracts, it is also true sometimes, arguably too often, that procurement priorities are then changed after the principal procurement path has been adopted.

All too often a client has said that programme is his priority, has then embarked on a path that puts it as the priority, only then to revert to saying that price must replace programme as the priority. Not surprisingly this can cause confusing effects on a design and construction team that does not then know which priority to follow. Naturally it is accepted that departure from the ingredients inherent in a procurement path may occasionally have to occur, principally if the procurement path itself was incorrectly chosen or if there is an essential reason for changing the inherent ingredients. This can usually only be done in the area of price and programme, not easily in the area of product. The quality of the product can be lessened by altering the price balance but, particularly once construction has started, it is not easy to increase quality except in superficial areas of finishes. The total quality of design cannot readily be changed. The balance of price and programme can sometimes be changed but very often attempts to reduce price, particularly after construction has started, required a slowdown in the programme so that more time is made available to consider issues of cost and price. A corollary is that a slower programme at this stage invariably increases some elements of the price, for instance, the contractor's site running costs will probably increase if the programme period is extended. A spiral of indecision and uncertainty over the priority of price or programme can then occur.

Personalities and responsibilities sometimes change between the choosing of a procurement path and implementing that path. This is often at the core of dissatisfaction over a particular procurement route on a particular job. The correctness of the information (PAC) that led to the PAO at the time of making the procurement choice, and its relevance to the events that actually occurred as the project proceeded through the design and construction phases, is an area where research may still be worthwhile.

The following case studies illustrate all procurement paths – three a management path, one a design and manage path, two a design and build path, two a develop and construct path and six a traditional path, either a sequential or accelerated path. A number of the cases have more than one procurement solution that would have been appropriate as a choice from the constituent parts of the study.

At the end of this chapter a list of the case studies is given alongside the PAO chosen for each of the projects. The procurement route eventually chosen is not given at the start of the case study and the order of the studies is deliberately varied between the procurement routes, so that the reader need not form any preconception on the procurement route chosen whilst reading the client, project and PAC/PAO details.

The majority of the case studies illustrate the choice of procurement to be a successful, mainly logical outcome of the client's priorities. In some cases the process is shown as not so successful or indeed so logical. Understandably the case studies illustrating relatively unsuccessful procurement choices have a longer text than those illustrating a relatively successful outcome. The postcripts have commented as appropriate on the relative success or failure of the PAO chosen.

The general lessons (see after the end of case study 14) summarise the overall observations on procurement and the central and crucial function of the client.

The case studies are:

1. Supermarket for Supermarkets Plc
2. City offices, shopping and leisure facilities for Citystyle
3. Hotel extensions for Retreat Hotels
4. Civic hall and theatre for Sometown Borough Council
5. Office for Growth Developments
6. Factory for Terrestrial Industries
7. Department store, unit shopping, offices and multi-storey car parking for Insurance Plc
8. Hospital for health authority

9. Prestige building for government agency
10. Shopping centre redevelopment for Central Plc
11. Clinic for Selfcare
12. Shopping centre renovation for Shopcentres
13. Golf country club for Bogey Leisure
14. Church rebuilding for St Dunstan's.

Case study 1: Supermarket

Client

Supermarkets Plc was among the small number of very large public companies serving the retail market, selling principally food and household goods. It operated over 300 stores and increasingly sought to control new store development and preferably to own its premises rather than rent them. It had begun recently to favour out-of-town sites rather than high street, in-town shopping locations. It had in-house executives solely devoted to new and renovation construction and was a major, frequent employer of construction consultancy and construction services.

Project particulars

At any one time Supermarkets Plc had under review perhaps twenty major locations for new store development. Quite often these were subject to lengthy planning discussions, sometimes to appeal, and sometimes they involved competition with other supermarket operators. Because of Supermarkets' programme of regular development, it used its own property and development department, together with consultants as required to obtain planning applications, negotiations and appeals.

For the purpose of this case study it has been assumed that the decisions on procurement advice had not needed to include any consideration of the need to build or any review of the location of the building and of the establishment of the size and function of the supermarket. All these matters would have arisen and been resolved from the continuous market research and internal company reviews that Supermarkets Plc carried out within their rolling programme of development.

The particular project particulars were for a supermarket of 200 000 square fee (18 600 m^2) with appropriate delivery facilities, customer and staff car parking and landscaping, situated on the outskirts of Oldtown. During the formulation of the brief within Supermarkets Plc, advice had been obtained from design consultants on the external appearance and massing of the building. This was necessary so that Oldtown Borough Council, which was the planning authority, was fully informed of the environmental impact that the supermarket development would have.

In this instance Oldtown Borough Council had received several applications and had therefore invited formal submissions from a number of retail food companies. This procedure provided competition for the sale price of the site that was owned by Oldtown, and enabled other issues, such as possible planning gain, to be considered.

Supermarkets Plc had therefore settled very broadly the type of building that they required, together with its general external appearance and the overall use they would make of the site.

Main considerations

The property and development department of Supermarkets Plc met, as part of their monthly review of major proposals, to review actions on the Oldtown project. The in-house project executive for the supermarket knew that outline planning permission for the project was likely to be granted but he did not know which retail company would be chosen to proceed to develop. He also knew that a decision was expected within two months and he decided to review his procurement arrangement options (PAO) accordingly. His main consideration was to be ready for fast development of the Oldtown supermarket if his company was the successful one.

Procurement assessment criteria (PAC)

Supermarkets Plc's in-house executive arranged a meeting of his financial, property and construction advisers, together with his external design and cost consultants.

A review of the PAC suggested:

- *programme completion* of the supermarket as early as possible was highly desirable. Average weekly turnover was projected at over £5 million and the amount of net profit was perhaps £150 000– £200 000 per week
- *significant variation* in the design, after commencement of construction, was not expected. Supermarkets Plc had commissioned over 200 new stores in the past six years and knew their requirements very well. Although they would need to respond to the latest retail requirements, variations should not be significant
- *complexity* was not considered significant. The building had air-conditioning and high security and safety controls but the technology was well-known. The construction was principally single-storey with some offices and welfare facilities on the first floor
- *the quality level* was naturally to be of good materials and workmanship.

The design of the building would follow the external concept already proposed in discussion and submissions made to Oldtown Borough Council. Design consultants previously employed by Supermarkets Plc would be familiar with the functional standards required

- *price certainty* was important but not critical. Construction costs were a very small part of the total considerations of Supermarkets Plc. Provided that the construction costs were fair value for money, resulting from efficient design and construction operations, price certainty within reason was not so important
- *price competition* was regarded as important. Retailers themselves were subject to competition and nearly always engaged and retained their own suppliers for their goods, with competition clearly demonstrated
- *management and accountability* were important for Supermarkets Plc. They had a large number of projects under way at most times and, although they had a very good in-house department, they preferred to delegate controlled responsibility to consultants and to contractors wherever possible. They favoured keeping things as simple as possible
- *risk avoidance/allocation* was important to Supermarkets Plc. Provided the risks were known in advance and where they should lie, that is, with consultants, with contractors and/or with Supermarkets Plc, a judgement would then be made on allocating them, in any contracts made.

In summary the PAC review established priorities as:

1. Programme – certainty and a fast design and construction period.
2. Price – competition should be demonstrated, but price certainty before starting construction was not important.
3. Variation and complexity were not significantly present.
4. Quality of construction product was important and quality of design was a factor, established from several previous developments.
5. Risk allocation was important.

Procurement arrangement options (PAO)

From the PAC review the following procurement arrangement options (PAO) were established:

(a) Management contracting/construction management
The size of the project was appropriate for management contracting as the project required a fast programme, with guaranteed completion as

perhaps the highest priority. Competition would also be obtained by management contracting.

(b) Traditional
If the programme could be 'guaranteed' and as early a start made with construction as possible, traditional may offer advantages in price certainty, whilst also demonstrating competition.

(c) Design and build, develop and construct, design and manage
Possible advantages that might be gained from total single-point responsibility were not possible in this case. The brief, the building concept and some of the design leading to an outline/scheme design had already been established in order to satisfy planning and other development requirements. The appointment of a design and build contractor could only be in a develop and construct role. Because virtually all the materials and detail design requirements were tried and tested parts of Supermarkets Plc's almost standard requirements, perhaps little scope would be allowed for interpretation by a contractor or would be required in the offer of a contractor's proposals.

Supermarkets Plc's in-house project executive asked his consultants, that is, the architect and quantity surveyor who had been involved so far in the project, to make any observations. The architect agreed that design and build/design and manage and so on was not appropriate. The quantity surveyor agreed with this and highlighted the lack of potential variations expected in the project and emphasised the relative simplicity of the construction. Although the project was in the medium price range the construction and engineering services were all relatively simple and had been constructed and installed relatively easily on other, similar projects. He highlighted that the key factor was that construction to a fast programme would depend on sufficient design being made available to a contractor, at the correct time.

Supermarkets Plc's executive asked his architect and quantity surveyor to consider, for a subsequent meeting, how the overall programme of design and construction could be made as short as possible, as this seemed the priority if permission to develop was obtained from Oldtown Borough Council.

At a subsequent meeting Supermarkets Plc's executive received the following information from his consultants:

- if full design and bills of quantities were required for tender purposes this process was expected to take approximately 7–9 months from receiving the instruction to proceed with design. This assumed using a design team already fully familiar with Supermarkets Plc's method of working

- a construction period of approximately 15–18 months was likely. This was consistent with experience on similar, out-of-town supermarkets
- a start on construction was only sensible if sufficient design information could be made available in the correct, regular sequence
- an earlier construction start could be made without the full design being completed at tender stage, provided foundation design and other early items that would be on the critical path for construction were to be made available on time. An earlier construction start should then mean an earlier construction completion
- either management contracting or traditional contracting, but using an accelerated route, would be appropriate for a programme of overlapping design and construction that was generally necessary for fast-track completion
- the expected design and cost control fees were estimated to lie in the range of 10 to 12 per cent of total construction cost – perhaps total design team fees would be around £1 500 000. The proportion of these fees that would be incurred up to tender stage would be very approximately 60 per cent of the total fee bill – perhaps therefore around £900 000. The amount incurred at any one point, along the path of preparing design and tender documentation, would naturally be only a proportion of the £900 000.

Supermarkets Plc considered the information they had now received on the design and construction programme, on consultancy fees, on the income and profit per week that they expected to make when the store would be trading and it was therefore decided:

- to proceed with consultant detailed design of the critical parts of the supermarket that would be required to start construction of the building, so starting the process to enable as early a construction start as possible. This meant the immediate start of architectural work and the engagement of structural and environmental engineering consultancies. Within six or seven weeks from when these appraisals took place, it was expected that a decision would be made on which of the supermarkets (from the three competitors) would be allowed to develop on Oldtown's site. Any abortive design fees incurred in the meantime, if the scheme were to be stopped if Supermarkets Plc were not the company chosen to construct and operate the supermarket, were considered to be a relatively small sum, if in return an earlier store opening date could be gained as a benefit if Supermarkets Plc were to become the chosen company.

    This decision was seen in relation to the considerable planning, design and public consultation expenses already incurred that would not all be recoverable from the project if Supermarkets Plc were unsuccessful

- the procurement route chosen was an accelerated traditional one, using drawings and bills of approximate quantities. Design by the consultancy team would progress in earnest as quickly as possible once the development was confirmed to Supermarkets Plc (otherwise design would naturally be stopped) and contractors would tender after approximately five months from that date on bills based on previous, similar projects, in addition to such design as had been developed by tender stage for the Oldtown supermarket. Foundations, steelwork, brickwork, engineering services and internal finishes would form the basis of the tender. Remeasurement of the construction as it had been build would establish the final sum to be paid.

Reasons for the procurement choice

The choice from the PAO was between either an accelerated traditional or a management contracting route. Design and build and design and manage variations, as discussed previously, were not appropriate.

Management contracting was not chosen because significant variation during the construction period was not expected to become a significant possibility (if this had been the case management contracting may have had a considerable advantage). Also the probable extra price that may have arisen from the management system was not thought worth paying. A traditional procurement route was thought to combine early competition with certainty over programme and a fair certainty over price. The feeling that there was the ability to allocate risk more substantially within a traditional contracting method was a significant point. The contract which was eventually let gave the contractor no extension of time for adverse weather or for his inability to obtain materials or labour, and it had a very high level of liquidated and ascertained damages. The design concept and all significant design was known to the contractors when tendering, and they all knew, and had previously worked for, Supermarkets Plc. They knew the programme, the product wanted and therefore the standards that were expected and that would be paid for in the price they contracted. The client was very experienced at purchasing the construction industry's services and products, knew what he wanted, insisted on getting it and would pay a fair price for it.

Postscript

Subsequently the supermarket was constructed, largely according to the procurement plan adopted. The procurement choice was very successful with the programme, price and product judged to be good.

Case study 2: City offices, shopping and leisure facilities

Client

Citystyle was a major development company, formed in the 1970s. Its chairman came from a property surveying, estate agency background and had developed his company to public status. Citystyle was an innovative developer and property company. It had developed business parks and large-scale commercial offices, often using innovative design from high-profile architects, to become one of the market leaders in prime locations. As well as developing these the company had begun to retain property as an investment and income source. Naturally Citystyle had developed by having in-house executives who knew their priorities very well.

Project particulars

Citystyle had conducted very extensive market research and interviewed likely clients and/or tenants to ascertain in considerable detail the development wanted by the prospective range of occupiers.

Citystyle had therefore assembled a site in a prime location in the City of London and were seeking planning permission to construct approximately 800 000 square feet (74 400m^2) of offices and 100 000 square feet (9300m^2) of shopping, speciality restaurants and leisure facilities, either for lease or for sale. The total development value was in excess of £725 million and the construction value was approximately £170 million. The offices needed to have a modern, prestige image to complement the location and the shopping had to attract people in an area that was then not highly residential. The provision of speciality food facilities, restaurants and perhaps some leisure facilities would be an innovation for the area.

Main considerations

Citystyle's development director had appointed an in-house executive whose sole responsibility was to arrange the construction of the offices, shopping and leisure facilities. The project was to be called Golden Gate, because it was on the fringe of the famous 'Square Mile'.

The development was to be carried out within Citystyle's overall aim in many of their developments, namely to have the latest construction start possible (in case the property market declined), to have the earliest possible construction finish (in order to minimise funding costs and lessen risk), to obtain rental income as soon as possible and to obtain the construction at the right price (to maximise the development margin).

From the beginning the in-house executive decided:

- prestige design of high-quality buildings was essential in order to attract tenants at rental levels (in the latter part of the 1980s) in excess of £50–£55 a square foot for offices and appropriate rentals to suit high-income shoppers and leisure-users
- because the market for such tenants and rentals was projected to remain for a number of years, but over-supply of offices in that location was projected within four to five years, an early completion and speed of completion was essential. The leisure facilities were seen as an attraction to large corporate clients, who from market research were expected or known to be the principal tenants or purchasers
- funding costs were high, partly because of the large size of the development. Interest rates, which were currently in historical terms relatively low, were expected to rise towards the end of the construction period, when most funding was required. This emphasised the need for a fast programme, both of design and of construction
- a proven team of designers and constructors was essential for the success of the project.

Procurement assessment criteria (PAC)

Citystyle's in-house executive arranged a meeting of his financial, property and construction advisers.

A review of the PAC suggested:

- *programme completion* was paramount. Expected rental income was good but the effect of a downturn in the property market required the risk of a long, uncertain programme to be minimised
- *significant variation* in the design after commencement was quite possible. Buildings of nearly 1 million square feet (92 000m²) would not be developed in one block or necessarily even in one phase. A single tenant, though highly desirable, may not be found (the office building(s) for instance would provide space for 6000–7000 people). Flexibility in responding to possible tenants or purchasers was necessary in such schemes. A fixed, immoveable internal layout would be a disadvantage. The space for leisure facilities would need to be flexibly provided
- *complexity* in the building form and in its environmental services was likely to be a factor. The external appearance was to be modern, prestigious, attractive to national or multinational organisations. Advice on this aspect and the importance of buildability that would allow fast construction was important

- *the quality level* was to be high, both in design and in execution
- *price certainty* was important but not critical. The construction cost for Golden Gate, excluding design fees, was expected to be around 25 to perhaps 30 per cent of the gross development value. Certainty in this area of the developer's equation was less important in this case than certainty of programme and product
- *price competition* was important. With a fast programme as a priority, competition was important wherever it could be used. Otherwise the effects of a fast programme may become distorted and too costly, if competition were not introduced at every opportunity
- *management and accountability* were to remain with Citystyle. Golden Gate was the biggest project so far for Citystyle and they would not let the reins of key decisions pass from them
- *risk avoidance/allocation* was important for Citystyle. Flexible response was the way Citystyle had learned to deal with risk, provided they knew in advance the areas where risk was most likely to be present.

In summary the PAC review confirmed priorities as:

1. Programme – this was the most important aspect.
2. Quality of design – this was essential.
3. The building was relatively complex – some innovative design elements may need to be incorporated.
4. Flexibility – this was needed to incorporate controlled variations if required during construction.
5. Price was less important – but only within agreed limits.

Procurement arrangement options (PAO)

From the PAC review the following procurement arrangement options (PAO) were established:

(a) Design and build, develop and construct
Citystyle did not consider this suitable. They were a client that was involved, even interfering, during the briefing, design and construction process and it was thought likely that the advantages of design and build (that is, of a clear brief established by the client in considerable detail before starting) would not occur. The quality of high-profile conceptual design was not thought to be readily available through a design and build organisation, and so would not meet Citystyle's wishes.

(b) Design and manage
Citystyle likewise dismissed design and manage, although appointment

of the whole team of designers and constructors through a consultant design and management organisation was considered.

(c) Traditional
The amount of overlap of design and construction required to meet the programme would not allow time for production of any form of bill of quantities that would be representative of the whole job, even by using an approximate bill. Two-stage tendering likewise was not appropriate.

(d) Management contracting/construction management
The very large size of Golden Gate was suitable for management contracting. A fast programme, with flexibility of response during construction in order to accommodate variations, was likely to be achieved by the employment of extra management resources, admittedly at a price, through adopting management contracting or construction management. Advice on construction procurement would be available, as a pre-construction activity, to enable purchasing lead times to be considered, critical materials to be programmed and advance orders placed.The procurement and buying knowledge of a major management contractor or construction manager would be essential to enable the design team to design with confidence, knowing their construction proposals could be purchased and installed in the building, in accordance with the overall programme required.

Citystyle then decided to make the following appointments:

- a construction adviser who would be responsible for pre-construction activities, including control of the overall design and construction programme. Advance orders for materials and early construction work would be placed through the construction adviser and he would become the management contractor or construction manager, subject to the satisfactory execution of the pre-commencement phase
- an architect and full design team with a reputation and adequate resources to produce a high-quality, fast-track design.

Reasons for the procurement choice

The choice from the PAO narrowed to the appointment of a high-grade design team and a contracting arrangement that allowed as much overlap of design and construction as possible. Construction advice during the pre-construction period, combined with the buying experience and power of a major management contractor, was seen as a good solution. The risk of slippage in the overall programme was the greatest concern to Citystyle and they regarded price certainty and, in some cases, competition as secondary concerns. Flexibility in the design and construction process had to be allowed for, to provide for some changes if the market required them. It was possible that, in response to the market, Golden Gate would be phased, accelerated or slowed as required. Only the flexibility of working within the framework of management contracting/construction management readily provided all these possibilities.

Postscript

Subsequently, a large proportion of the development was constructed, broadly to the programme adopted, before a change in the property market caused development to slow down. The procurement choice was very successful and the client, consultants and contractors very broadly met and were satisfied with their objectives.

Case study 3: Hotel extensions

Client

Retreat Hotels was a private company that specialised in the operation of hotels generally situated in pleasant rural or out-of-town settings. It catered for a part of the hotel market that liked an individual hotel, probably in an older building, situated in a country district. Restaurant facilities were required to match this market. Retreat had operated for four years and now owned three hotels, each of 20 to 30 bedrooms.

Hotel market research consultants had advised Retreat Hotels that, if they wished to expand their operation and increase the number of hotels they owned or operated, a significantly large market existed for the type of hotel that they provided. Retreat Hotels therefore decided to obtain private financing, via a merchant bank, and to expand their operation rapidly over a period of two to four years by acquiring approximately 10–14 hotels and, more importantly, to increase the size of their hotels where possible. Both newly acquired hotels and also existing operations would be enlarged, so that each hotel provided 40 to 60 bedrooms, together with leisure, conference, improved restaurant and non-residential facilities.

Project particulars

Retreat Hotels, with advice from their hotel market research consultants, purchased three hotels in their first year of operation. Before purchase of each hotel Retreat established that, in principle, the planning authority would allow an increase in size of the hotel, along the lines of Retreat's general policy. At the same time Retreat made planning applications for their existing hotels so that they could extend them in a similar way if required. In order to apply for planning permission Retreat had used the services of one or more architects, who had in some cases prepared outline schemes for the previous owners of the hotels that had recently been acquired by Retreat. Retreat did not have an in-house executive who was either experienced enough or able to devote sufficient time to the expansion of their properties. Their managing director therefore sought advice from their merchant bank and it was agreed to take advice initially from a quantity surveyor on the appropriate procurement route(s) to adopt. Retreat had previously engaged a quantity surveyor to help in the settlement of one of their building contracts for minor works, a contract that had run into difficulties. Surveying Partnership were therefore asked by Retreat to advise on the expansion of their hotels.

Main considerations

The managing director of Retreat met the senior partner of Surveying Partnership and told him that, in his view, the following were the main considerations:

- Retreat Hotels had not been used to dealing with construction work
- they were concerned about building delays, extended programmes and overrun of costs beyond their budgeted allowances. The experience they had in dealing with a small local builder had not been pleasant for them
- they must provide a good service to their hotel guests during any extension and/or alteration work to any hotel
- certainty of price was important. They wanted to know how much any works would cost before deciding to start a development
- completion to an agreed programme was also important, bearing in mind commitments to seasonal functions such as Christmas, main holiday peak seasons and the like.

Procurement assessment criteria (PAC)

Surveying Partnership explained to Retreat's managing director the options available to him, using simple terms from the NEDO guide *Thinking about Building*.

At the end of the review Surveying Partnership and Retreat agreed:

- *programme completion* of the extension of all Retreat's hotels was not critical. The decision to expand the whole of Retreat's operations had been taken involving a sensible, measured programme for each hotel as appropriate
- *significant variation* in the design of extension and alteration works was not expected after commencement of construction works. Retreat's managing director explained that the operating policy of a hotel or of a hotel group was a matter of philosophy arising from the owner's preference – one hotelier would perhaps do certain things somewhat differently from another hotelier. Having said that, the managing director expected that a decision on the extra facilities at each Retreat hotel would be made, then planning permission obtained, and construction work carried out to the agreed plan. Variations were expected to be relatively few once construction had started
- *complexity* was not considered significant. The building construction required was generally two- or three-storey, in the manner of a large country house. Environmental services were generally simple with comfort air-conditioning provided only to conference/meeting room facilities and leisure facilities
- *the quality level* needed to be good. Extension works would have to match an existing hotel and sometimes the extension may be larger than the original hotel. Sympathetic treatment of the external appearance was important to maintain the image of Retreat
- *price certainty* was most important. If the price of extending a hotel was too much the hotel would not be profitable and therefore would not be extended. The final price needed to be known as nearly as was possible, at the outset
- *price competition* was needed. Competition was the way to obtain the market price, based on clear, known requirements
- *management and accountability* were important for Retreat. They did not have the resources to manage separate consultancies and contractors. They favoured single-point responsibility for as many services as possible
- *risk avoidance/allocation* was most important. Retreat would prefer to put as much risk as possible with a single organisation. They realised this may possibly mean a higher price.

In summary the PAC review established priorities as:

1. Price – competition leading to price certainty at contract stage was the most important matter.
2. Programme – once works had started, because of the need to maintain operation of the hotel, any closedown or restricted operation

must be to a known programme and completion must be on time.
3. Variation and complexity – these were not expected.
4. Good quality work – this was essential
5. Risks – and the penalty for failure should be with the contractor as much as possible

Procurement arrangement options (PAO)

From the PAC review the following procurement arrangement options (PAO) were established:

(a) Management contracting/construction management
The likely value of the works, about £2–3 million per hotel on average, made management contracting inappropriate. Variation during the works was not expected and management contracting would not offer price certainty as defined by Retreat.

(b) Traditional
The works needed to be designed to outline stage by an architect in order to obtain detailed planning consent. Full design could then be carried out by an architect, structural and engineering consultancies and a tender obtained, probably using a bill of quantities. More consultancy fees would have been involved in this method if a project were to stop at tender stage. Price certainty should be possible by this method. A split in responsibility between consultant designers and the contractor was involved. Instructions issued by the architect may have become a source of delay and/or extra costs.

(c) Design and build, develop and construct, design and manage
The single-point responsibility of these options appealed to Retreat. Design and manage was discarded because it would not provide price certainty. Design and build was not applicable because the planning permission would largely prescribe the layout and major elements of the building form. Develop and construct would enable a contractor to develop a prescribed design within certain limits, to offer a price and programme with reasonable certainty and to provide single-point responsibility to Retreat.

Following the review of PAO, Retreat Hotels and Surveying Partnership met the architects appointed to obtain planning permissions for the three hotels purchased that year, and also those acting where permissions had been applied for on the three hotels already owned by Retreat. The PAO were discussed with these architects and their preference and opinions on procurement methods were obtained.

After this review period, which lasted about three months to consider all six hotels, it was decided:

- full planning permission should be obtained for extension and alteration works to six hotels
- architects should be commissioned to produce a scheme design for the extension and alteration works involved in each of the six hotels
- develop and construct tenders should be obtained for the six hotels
- an 'employer's agent' would be appointed for each project and he would prepare 'employer's requirements' for each project, based on the design prepared to obtain planning consent
- tenders for each of the hotels would be obtained from contractors (some of whom may be appointed to construct more than one of the hotels, if appropriate) on a develop and construct basis using a standard form of design and build contract. The 'contractor's proposals' for each hotel would need to be carefully matched with the 'employer's requirements' before any contract was signed.

Subsequently Surveying Partnership became the 'employer's agent' on four of the projects and a firm of architects was appointed for this role on the other two projects.

Reasons for the procurement choice

The choices available from the PAO were between the appointment of a traditional contractor, based on a bill of quantities tender with an architect's and other consultant's full design drawings (possibly a specification, without quantities tender was a variant of full design but £2 million without quantities was not perhaps appropriate) and the appointment of a contractor on a develop and construct basis. In this arrangement detailed 'employer's requirements' would need to be given to the contractors tendering, including programme constraints, some constraints in the materials and workmanship to be used and methods of construction that the contractor must provide. In other respects the develop and construct contractor would be free to put forward, in his 'contractor's proposals', any construction methods he wished to adopt.

The appeal to Retreat Hotels of the develop and construct option was that they would obtain the architectural concept that they wanted by obtaining detailed planning consent by having an architect take his design up to scheme design stage. Retreat would then place single-point responsibility with one organisation for the development and implementation of the architect's design, for materials and workmanship and for the construction programme, thus passing major risks to one organisation. They should obtain a high degree of price certainty before entering into a construction contract. They would reduce the amount of consultancy fees that they would incur in their own professional team by going down the develop and construct route, as opposed to the traditional full design route, if for whatever reason a project did not go ahead after tender stage.

Postscript

Subsequently, a number of hotels were developed according to this procurement plan. The amount of variations that occurred after construction had started meant that both programme and price were not the certain things that were hoped for at contract stage. Because the 'employer's requirements' were not thoroughly worked out at contract stage and/or were not rigidly adhered to, significant variation occurred with resultant adjustments to programme and final price. However no other procurement option would have been more appropriate.

Case study 4: Civic hall and theatre

Client

Sometown Borough Council was a local authority with a population of approximately 350 000. It was responsible, through a Technical Services Director, for building facilities, capital projects and their maintenance. As Sometown had expanded it had needed to erect multi-storey car parks, social service facilities, some public housing and ancillary services but it had not been responsible for building projects of any size or complexity.

Project particulars

Sometown Borough Council decided to provide a civic hall and theatre in their town centre, within an area that was planned to be redeveloped by a comprehensive shopping and office scheme. The council formed a sub-committee that reported to its Policy and Resources Committee, in order to look at the development of proposals for the civic hall and theatre. The subcommittee was advised by the Technical Services Director of the council. The civic hall was budgeted at £7 million and the theatre at £15 million, both at current prices.

At the first subcommittee meeting, when a number of councillors made proposals ranging from appointing architects, holding an architectural competition, appointing project managers, appointing design and build contractors and so on, it was agreed that some more basic advice should be obtained. It was decided that a local quantity surveying consultancy, that had provided quantity surveying services to the council for a number of years, should be asked to advise the Technical Services Director, who would then report to the council, on all aspects of the proposed theatre. Accordingly QS Consultancy and Partners, of High Street, Sometown, were appointed as Sometown's 'principal adviser' for the proposed civic hall and theatre. At that stage their commission was only to provide procurement advice.

QS Consultancy and Partners (QSCP) met the Technical Services Director and discussed very broadly the need for the hall and theatre and how QSCP would work with Sometown Council. The council would need regular progress reports and the Technical Services Director appointed his deputy as Sometown's in-house project executive. The deputy would report to the director who would in turn report to the subcommittee of the Policy and Resources Committee, who in turn would report to the full council. It was agreed that the QSCP 'principal adviser' would meet the deputy director weekly, so forming a project group, set up to establish construction procurement conditions. Full delegated powers, consistent with periodic reporting to the subcommittee of Policy and Resources, were given to the deputy director.

Main considerations

At the first meeting between QSCP and the deputy director it was agreed:

- that the budgets allowed for the hall and theatre should be reviewed. QSCP agreed to provide historic costs for similar projects. The deputy director agreed to confirm the funding arrangements for the project
- that the programme should be established as soon as possible. Statements had appeared in the local press stating that Sometown Council hoped to have the hall and theatre open in three to four years' time when a gala year of celebrations would take place with a 'twinned' German town. Privately QSCP and the deputy director thought this programme optimistic and they wished the constraint had not been suggested
- that the location of the project was to be confirmed. A central area redevelopment was to be carried out by a property company in Sometown and the civic hall and theatre were planned to be adjacent.

Procurement assessment criteria (PAC)

QSCP and the deputy director met weekly and after eight weeks agreed that the PAC were:

- *programme completion* of the hall and theatre in a period of three to four years, as requested, was desirable but looked difficult to achieve. The briefmaking process for the civic hall would take approximately three months. The briefmaking process, needed to establish the facilities of the theatre, would take approximately six months and would probably require advice from a theatrical management consultant who was yet to be appointed. An architect would need to establish the two briefs, prepare the client's requirements and then prepare sketch proposals for the civic hall and theatre. Perhaps some consultation process with Sometown's inhabitants also needed to be considered. If the initial aim of overall completion in three to four years were to remain, some overlap of the design and construction process seemed to be necessary
- *significant variation* was not wanted. If the brief and design were properly established then changes should not take place
- *complexity* may be a factor. The multi-purpose hall was relatively simple but the theatre was not. Air-conditioning would be required and the electrical installations may need specialist advice
- *the quality level* of the project was to be high but not extravagant/prestigious. There was often a delicate balance to be maintained in civic construction when public money was involved. Insufficient and/or inappropriate design and cost consultancy skills existed in the Sometown Technical Services Department. Therefore consultancy appointments would be required
- *price certainty* was usually a requirement in local government projects and it was expected to be so with the hall and theatre. Some of the councillors had quoted 'value for money'. Competition and control of expenditure within an agreed budget would be easier if price certainty at the start of construction were possible. Usual procedures for applying for approvals and monitoring and reporting would apply
- *price competition* was important. Correct expenditure of public money could be demonstrated most easily by competition
- *management and accountability* were important. Sometown did not have a large in-house Technical Services Department and it would like to delegate controlled responsibility to others
- *risk avoidance/allocation* was important. If the risks could be identified Sometown would decide where to allocate them

In summary the PAC review suggested priorities as:

1. Programme – the overall period was critical.
2. Price – the budget was not ample and certainty of price before construction started was important.
3. Complexity/quality – the project was not simple construction and the design must be sensitive.
4. Risk – allocation was unclear at present.

The period between the appointment of QSCP as the 'principal adviser' and the review of PAC leading to PAO had, because of consultations within the Sometown Council, taken over three months. Within an overall programme of three to four years this was too long a time to spend before other important decisions and actions needed to be taken.

Procurement arrangement options (PAO)

From the PAC review the following procurement arrangement options (PAO) were established:

(a) Management contracting
Sometown had no experience of this procurement route. The Policy and Resources Committee had among its members a property development consultant and a building contractor. Their views were not generally in favour of the management route.

(b) Traditional
The programme required, of completion within three to four years, would be difficult to obtain if the usual sequential path of traditional procedures were to be followed. If a traditional path was thought to be appropriate, accelerated procedures would be required.

(c) Design and build
The design quality required for the hall, and more particularly for the theatre, did not suggest that a design and build route would be appropriate. The client's briefing process required work still to be done to establish spatial requirements and relationships. Once done this could have been converted into 'employer's requirements' and then design and build tenders sought. However, it was felt that the quality of external design, obtainable by this route, would be inappropriate to Sometown's requirements.

Sometown decided to make the following appointments:

- QSCP were to remain as 'the principal adviser' and they would also be appointed to provide cost consultancy services
- a firm of local architects, together with other consultants appointed when required, were to commence briefing and then to prepare outline design for the hall and theatre
- a firm of theatre management consultants were to advise on the operational aspects of theatre management. The operational policy adopted by Sometown for the theatre could affect the design of the theatre.

It was decided the procurement route would be an accelerated traditional one, using approximate documentation to obtain earlier contractor appointment than a traditional sequential route would allow. The overall programme for design and consultation was still unsure and appointment of a contractor did not seem to be required for some time.

Reasons for the procurement choice

Sometown's experience was limited to traditional procurement methods, and latterly to some limited use of design and build. Sometown was concerned to show price competition and to obtain price certainty, whilst attempting to have the civil hall and theatre completed to their programme. The council felt that management contracting would not provide enough evidence of competition and would result in uncertainty over the cost of the project before they had started construction.

Postscript

The project briefing for the civic hall was completed and a scheme design produced after approximately twelve months from the architect's appointment. The theatre briefing and design stages took approximately 18 months. Since the decision to have a hall and theatre was taken, approximately 21 months of a 48-month programme had elapsed.
QSCP and the Sometown Director of Technical Services reviewed their procurement options again at this stage, when 27 months were left of the original programme, if the facilities were to be completed in time for the year of celebration with a twinned German town. The earlier intention to follow a traditional accelerated path was again confirmed in lieu of a management path. A contractor was appointed following a tender based on approximate quantities. Throughout the course of the contract the contractor was concerned at the amount of design

changes and he requested several extensions of time, finally finishing the hall and theatre 15 months after his contract date. The final cost was considerably over budget and a number of contract disputes took several years to resolve, with all involved incurring significant loss and/ or expense. The hall and theatre were not finished in time to be used for events during the year of 'twinned German town' celebrations.

In retrospect the original requirements for the design and construction programme of three to four years should have been challenged and explored much more at the outset. Time that was allowed to be taken in the early stages of design was insufficiently controlled and an attempt was then made to make up this lost time by embarking on an accelerated traditional route. This was started with insufficient design information available at contract stage. At this point, had the programme really remained the priority, a management route should have been considered more carefully. The choice of the traditional route, because the contractor was appointed on the basis of incomplete design, led to price uncertainty combined with programme uncertainty. The building contract gave no incentive for the contractor to adapt his programme or provide extra management resources to bring the project back on programme, if possible. The client, consultants and contractors were dissatisfied with the project although the completed project is now functioning well.

Case study 5: Offices

Client

Growth Developments was a private development company that had existed for six years. It was owned by three of its directors together with a major share-holding by a bank. It had started operation in small residential developments, then it transfered into light industrial and latterly became interested in office developments.

The chairman and the managing director approved all investment and procurement decisions, but on occasions the bank (the majority shareholder) requested consultation if a project was over £8 million in construction value. Growth generally pre-sold their developments to an insurance company that placed some of its investment income directly into property purchase. In this way Growth did not tie up their resources in property ownership. Perhaps this would come at a later stage in their development.

Growth knew a number of consultants and contractors who had worked with them over the past six years.

Project particulars

Growth Developments had under review a site in one of the expanded towns along the M3 motorway. They had been advised by their London property agents that the area was one of growth for office employment and that the office rental levels justified a certain construction cost, so that Growth could develop and then sell (before but possibly during building construction). The site allowed a development of perhaps 90 000 square feet (8 370m²) of offices within an overall 'business park' setting. Landscaping and proper attention to the general location was important. The construction value was in the order of £8 million.

Main considerations

The proposed development was the largest that Growth had undertaken so far and it was agreed that their managing director would become the in-house executive responsible for the development. The size and risk of the project was his principal concern as he approached his first detailed discussions.

Procurement assessment criteria (PAC)

Growth Development's managing director and a senior partner of Growth's property agents, who were advising on the purchase and disposal of the site, met to establish the following PAC:

- *programme completion* was important. Although the development may be pre-sold, that is, sold before construction was completed, to an insurance company, they in turn would expect to obtain a rental income at the time that they had planned
- *significant variation* in the design should be minimal once construction started. Growth's managing director, who came from a construction background, saw no reason to allow changes in a speculative office development, unless a prospective tenant agreed to pay for them
- *complexity* of building was not involved
- *the quality level* was good but nothing special
- *price certainty* was important. Growth wanted to know the price before they agreed to proceed. Their development margin would then be known, subject to the cost of borrowing money and the sale costs of the completed development
- *price competition* was important. Growth firmly believed that the price to them was lowest when competition was used
- *management and accountability* were important. They liked things

to be as simple as possible and, generally, they felt that more people's involvement made for more complications

- *risk avoidance/allocation* summarised all the above for Growth. They wanted to take a development margin from providing a product, on time, to a known price. At present they were not interested in taking a risk on future rental levels, inflation in construction tenders and associated matters to any large extent.

In summary the PAC review established priorities as:

1. Programme – certainty and a fast design and construction period.
2. Price – competition and price certainty before any construction was started.
3. Variation and complexity were not factors.
4. Responsibility and risk – these were to be minimised. Growth wanted to place both on to someone else's shoulders.

Procurement arrangement options (PAO)

From the PAC review the following procurement arrangement options (PAO) were established:

(a) Traditional and management routes
These were not considered to be applicable. Traditional would take longer than wanted by Growth and would involve payment of consultancy design fees, at risk to Growth, if the project had to be stopped. Separation of designers from constructors did not appeal to the managing director of Growth, unless overall quality of design required it. Management contracting would provide a fast completion programme but gave less price certainty and split responsibility for design from that for construction.

(b) Design and build, develop and construct, design and manage routes
These routes seemed very appropriate because they could offer price and programme certainty and competition in all areas of design, as well as in construction. Minimum risk to Growth would be present if tenders were obtained and development proposals had to be stopped at tender stage.

Growth decided to obtain competitive tenders from design and build organisations.

Reasons for the procurement choice

Growth Developments wanted price certainty and price competition more than programme certainty. They expected, once they had set up the land purchase and obtained a building price, to sell the development to an insurance company. Although they recognised that the overall programme for design and construction could probably have been shortened by the use of management contracting or by an accelerated traditional route, they did not want the risk and relative uncertainty of those procurement routes.

Subsequently Growth produced, with the aid of an architectural consultant, a statement of 'employer's requirements', giving the space and quality standards that they wanted. Very little constraint was placed on the design and build tenderers concerning height and configuration of building(s) required.

Design and build contractors that had in-house design departments and who, when appropriate, based their work on a structural system, probably one patented by them, were invited to tender. This category of design and build tenderer was chosen because the overall time spent in their design, before construction could start, was expected to be reduced because of the use of standard, tested, structural solutions. As a result of this, it was also expected that the design fees would be lower than by inviting design and build contractors who would then have to engage consultant designers.

When tenders were received the evaluation of them was relatively difficult, because the 'employer's requirements' deliberately did not constrain enough matters. Comparability of offers was not evident on the face of the tenders received. This is sometimes a difficulty inherent in design and build tendering. A 'second stage' tender process was required during which two of the original tenderers were invited to amend their proposals, following further employer's clarification. The NJCC *Code of Procedure for Selective Tendering for Design and Build* is a useful reference for good practice in this area.

Postscript

The offices were constructed according to the contract programme and price. Quality was consistent with the product offered and the procurement choice was successful.

Case study 6: Factory

Client

Terrestrial Industries was a public company involved in many activities ranging over manufacturing, distribution, wholesale, retail and communications. They had offices, research establishments, factories and facilities in many countries, including the UK. Their UK origins meant that their procurement policies, at least in the UK, were still influenced by UK procurement methods.

Project particulars

Terrestrial wished to expand the manufacture, storage and distribution facilities of electronic components, mainly used in high-technology defence and aviation industries, but increasingly used in consumer goods. Price competition from overseas imports was expected to make Terrestrial's product unfavourable within five or six years, when Terrestrial then planned to have started production of the second generation of components. They planned to leave importers to take over the supply of the first-generation components whilst Terrestrial then concentrated on the expected lead they would by then have gained in the second-generation product.

Terrestrial invested very large sums of money in 'facilities', as buildings were collectively known. The location of a facility was determined by availability of government grants, good or bad labour relations, transport communications and so on. Many matters not directly related to building were explored with property consultants and financial and production analysts before a site was finally chosen. Feasibility studies for several locations were simultaneously carried out in areas ranging from western Scotland to the north-east of England, to South Wales and the Midlands.

The eventual site was chosen principally because of its location, adjoining two motorways, and because of the investment grants available in the area – a large amount of the investment would be funded by investment grants and also by investment allowances.

Terrestrial had, throughout the feasibility studies of different locations, already considered 'building facility procurement' and they now reviewed their initial ideas. The construction value of the facilities was in the order of £40 million.

Main considerations

Terrestrial's in-house executive for 'project X', as it was code-named, knew the following factors needed to be taken into account:

- the market period for Terrestrial's components was limited, because imports would eventually enter the country at a lower, unbeatable price
- the technology of making, storing and handling the new product required a controlled, sophisticated environment in the facility.

Procurement assessment criteria (PAC)

A review of the PAC established:

- *early completion* of the facility was essential
- *variation* in the design, during construction, was very probable when Terrestrial established, from their development and marketing departments, the exact requirements for their new product, and therefore the facility in which to make it
- *a complex,* environmentally sophisticated facility was required. 'Clean air' was synonymous with advanced technology
- *the quality level* of the facility was not inherently important. Within five or six years it may have become of limited or of no economic use, at least to Terrestrial
- *price certainty* of the facility was relatively inconsequential
- *price competition* likewise was relatively unimportant
- *management, accountability and risk avoidance* were important. Terrestrial wanted to know exactly what risks were inherent in any procurement, where they lay and what were the alternatives. They expected to be involved with the production of the new facility and would not be taking a 'hands off' role in its design and construction.

The PAC review confirmed priorities as:

1. Programme – guaranteed production facilities on time.
2. Product quality – the new facility must reliably allow the new electronic components to be continuously produced.
3. Complexity of facility – the design and construction of the facility could only be carried out to the programme required by an integrated organisation, familiar with sophisticated environmental conditions.

Allstores decided that the Princetown department store was not efficient because it was too large. The store had to be reduced in size and the surplus space thereby made available would be able to provide individual shop units, separate from but adjoining the Allstores department store. A multi-storey car park and some offices would also be developed. The amount of construction work involved was estimated to be in range of £75–£100 million depending on the extent of construction. During any rebuilding/renovation work Allstores wished to continue 'business as usual' as far as practicable – a total closure of the store did not make financial sense.

The Allstores in-house project executive had carried out several reviews with his trading, marketing and property departments and concluded that in principle approximately one-third of the existing store area would need to be released to provide new shopping units, after the remaining two-thirds of the existing department store area had been rebuilt and/or renovated to produce a new department store.

In reviews of a number of possibilities, Allstores had used one design practice, from among a number of design practices they regularly used for their retail work. However Allstores were not content to proceed with such a major redevelopment on their own and they decided to approach a number of property companies to obtain competitive submissions from them. In this way they obtained a range of financial offers for redeveloping the whole of the area occupied by the existing Princetown store. The range of offers received was from one of purchase of the property from Allstores and then leasing back a redeveloped store to them, to one of taking a lease from Allstores and then redeveloping the property on their behalf. Several combinations of offer involving payments of a capital sum to Allstores, change of ownership, joint ownership and no change of ownership were involved.

For the purpose of this case study it is not necessary to detail the process of seeking, evaluating and accepting offers from property companies and investors. It is sufficient to say here that six property companies were invited to make submissions based very broadly on the Allstores requirement to increase the efficiency of the area that they used for trading as a department store, so releasing the space left over to become unit shops, offices, car parking or whatever may be allowed by the planning authority. Major design practices and construction organisations were engaged by the six property companies to develop their proposals. Models, many drawings, financial appraisals, estimated construction costs, programmes and associated matters were produced. Evaluation of the offers was a complex operation. The process took approximately 18 months and went through several stages, finally involving two of the original six property companies.

Allstores chose an offer from Insurance Plc who became, for the purpose of this case study, the client for procurement of construction.

Main considerations

Insurance Plc negotiated an arrangement of providing a renovated department store for Allstores, shopping units for rental by separate retailers, office space for tenants and multi-storey car parking for public use.

Certain parts of the scheme needed to be developed quickly whereas other parts were less urgent. Insurance Plc had worked, intermittently, in competition with the five other property companies for nearly two years on a series of schemes for redeveloping Allstores' property. During this time they had used a full design and cost consultancy team and, once appointed by Allstores, Insurance Plc invited the team to a meeting to discuss procurement options.

Insurance Plc's in-house executive (a senior property director) told his consultancy team that the following were the main considerations:

- the total development must be carried out in phases with the main department store redeveloped as soon as possible to suit the agreement Insurance had made with Allstores.
  The phases would be:
  Phase I – renovation of part of the department stores and construction of new space for the remainder of the store
  Phase II – construction of new multi-storey car parking
  Phase III – construction of new shopping units
  Phase IV – construction of new offices
- the shopping units could not be constructed until part of the area of the existing department store became available, following renovation/construction of some of the new space for Allstores' department store
- office and multi-storey car park facilities were not on a critical path
- a good quality of design was necessary and overall the total development should be carried out within five years
- a proven team of designers and constructors was essential for the success of the project.

Procurement assessment criteria (PAC)

A review of the PAC suggested:

- *programme completion* was important particularly in the first phase, less so in the subsequent phases. Because of phasing, a single procurement route may not be appropriate for all phases

- *significant variation* in design after commencement of construction was not wanted by Insurance Plc. Perhaps in phase I (the department store) some variation was possible, because of the nature of renovation combined with new construction. Subsequent phases should not be subject to variation because they were to be mainly of new construction
- *complexity* in the building form of phase I was involved. The renovation/new construction of the space to provide a smaller department store included an atrium, several escalators, air-conditioning and maintenance of an external facade to the periphery of the existing store. The construction was relatively complex and alteration and new construction had to be carried out partly whilst 'business as usual' trading continued in as much of the department store as possible. The other three phases were not particularly complex
- *the quality level* was to be high both in design and in execution for the department store, less so for the other phases
- *price certainty* was important but not critical. The assessed gross development value meant that some contingency was allowed for the unknowns of construction costs, within limits
- *price competition* was important. Phase I may not allow this because programme was critical but subsequent phases should be competitive wherever possible
- *management and accountability* were to remain with Insurance Plc who would be an involved client throughout all stages of design and construction
- *risk avoidance/allocation* was important for Insurance Plc and they would prefer to pay for devolving risk to another organisation wherever possible.

In summary the PAC review confirmed priorities as:

1. Programme – this was important for phase I but less so for the other three phases.
2. Significant variation – this may be unavoidable in renovating parts of phase I but it should not occur in principle in the other phases. Because more design time would be available for phases II, III and IV it was planned that significant variation would not be required or allowed to happen once construction had started.
3. Complexity – this again was present only in the renovation/new construction of the department store.
4. Price certainty – this was particularly required once the subsequent phases were procured.
5. Price competition – this was required more in the subsequent phases but also wherever possible in phase I.

Procurement arrangement options (PAO)

From the PAC review the following procurement arrangement options (PAO) were established:

(a) Design and build, develop and construct
Insurance Plc did not consider this was suitable for phase I, the renovation of the department store. It may only be appropriate for obtaining the multi-storey car parking.

(b) Design and manage
This was not considered appropriate.

(c) Traditional
The amount of overlap of design and construction required for the store renovation would not allow enough time for a full design to be prepared and then a firm bill of quantities to be produced for phase I. The other phases would be designed during the design and construction of phase I and their procurement could be obtained using traditional routes.

(d) Management contracting/construction management
The size of the total project, at £75–£100 million, and the programme requirements were suitable for management contracting, particularly for phase I. (Phase I value was approximately £40 million.) Some advice on construction methods may also have been required for renovation/rebuilding of phase I.

Insurance Plc then decided to make the following appointments:

For phase I – renovation and rebuilding of the department store:

- a management contractor who would be responsible for pre-construction activities, including management/control of the overall design and construction programme. This would enable advance orders to be placed through the management contractor. A considerable amount of work to the existing department store would have to be carried out on a 'prime cost' basis, organised by the management contractor, some of this work perhaps being done during the period of pre-construction activities. The work required was to demolish and rebuild sections, to provide extensive temporary screens, temporary public accesses and routing. This would be required in order to maintain the trading activities of the department store during the time it was prepared for reduction in size by approximately one-third. 'Prime cost', arguably a form of management contract but without many elements of competition, was appropriate to this situation as it allowed flexibility of labour resources to be applied as and when the work and programme required. It was recognised that it could be a very expensive way of working but, particularly when dealing with alterations to and within an existing retail store that must continue to trade, and when the extent of the work involved could not be entirely foreseen, it was an appropriate contract arrangement

- an architect and full design team capable of fast-track design.

For phase II – construction of multi-storey car park:

- a design team to produce full design and tender documentation leading to a traditional contract
- appointment of a contractor following selective competitive tendering.

For Phase III – construction of new shopping units:

- a design team to produce partial documentation leading to a management route
- appointment of a management contractor. Negotiation or competition with the contractor appointed for phase I would take place.

For phase IV – construction of new offices:

- a design team to produce full documentation leading to a traditional contract
- appointment of a contractor following selective competitive tendering.

Reasons for the procurement choice

The choice from the PAO required a high-grade design team and contracting arrangements to suit each phase. Construction and programme advice on phases I (new store) and III (shopping units) would be needed as these would form a single entity when completed. It was thought appropriate to use the flexibility of management contracting because design could not be completed before construction had to start. The priority of completion of construction for phase I was most important in order to allow the commencement of construction for phase III.

Phase II (the multi-storey car park) was a construction on its own, not on a critical path for the overall development. Competition and price certainty were best achieved by producing a full design and using traditional competitive tendering.

Phase IV (new offices) was not interlinked by design with the shopping or the car parking and the overall programme for the whole site allowed sufficient time for a full design and traditional competitive tendering.

Postscript

Subsequently a large part of the development was constructed, including the car park. The office content later became subject to review and immediate development was not commenced. The procurement choice for phase I was correct with the programme that had to be met but price control was not obtained. Phases II and III had programmes that allowed traditional processes to be followed, but it was felt appropriate for phase III to continue from phase I along a management route but with better price control.

Case study 8: Hospital

Client

Regional Health Authority (RHA) was responsible for the provision of health care within its region as part of the UK National Health Service. As part of this responsibility RHA had to provide building facilities, including hospitals, from time to time. RHA had to be fully conversant with the procedures for design, cost control and overall approval in sequential stages as contained within the National Health process.

Project particulars

A 300-bed hospital was to be provided in Healthtown. This town had been chosen after many exercises involving optional towns, with different facilities and considerable combinations of new and renovated construction. Perhaps nearly six years had passed during which this intermittent process took place.

RHA policy for major hospital design was to commission a full design team and then to obtain competitive tenders for construction.

Main considerations

RHA wished to obtain authority from the Department of Health by following well-known procedures. Capital budgeting and control and allocation of resources within that process was essential to the operation of RHA.

Procurement assessment criteria (PAC) and procurement arrangement options (PAO)

The processes referred to above meant that only in exceptional circumstances was a formal review of PAC necessary. RHA appointed an in-house executive for a proposed hospital and he administered Department of Health procedures. The most important PAC was to have a consultant design completed at tender stage such that a certificate of readiness to proceed was obtained from an architect. Provided that tenders were obtained within an agreed budget then price certainty criteria at contract stage were satisfied.

The PAC of competition was obtained by competitive tenders and the PAC of controllable variation should be satisfied by the 'full design' requirement. The PAC of programme, complexity and quality level were not significant for RHA. The PAC of responsibility and risk avoidance were satisfied for RHA, as they judged these criteria.

Reasons for the procurement choice

Because RHA was subject to capital cost control procedures, and hospital designs generally must satisfy well-known functional and spatial requirements, the process of PAC review was very simple. Programme, that is so often a priority in commercial activities, was not a high priority in the provision of National Health care facilities. A case for using management contracting could not be made and one for using design and build was hard to make, if consultant design were to be taken up to a full design stage. Unless the PAC had circumstances that had put programme as a priority, management contracting did not seem appropriate because its advantages would be mainly in responding to unplanned variations and in having contractually flexible programming abilities. Likewise, whereas design and build in a total sense precluded consultant design, a case could have been made, in some circumstances, for a develop and construct procurement path, after a 'scope design' had been prepared by a consultant.

Postscript

The procurement path was successful within its constraint of little concern with overall construction programme certainty. A considerable number of variations did occur for a number of reasons, although full consultant design was apparently completed at tender stage. Delays occurred and the contract period was considerably extended with associated effects on the final price.

Case study 9: Prestige building for government agency

Client

Alltech Agency (AA) was established during the 1980s to procure government buildings for defence purposes. Previously other government agencies such as the Property Services Agency provided basically the same service. It had been decided that better procurement of many things from weapons to paper clips could be obtained by using AA. It had become a very large procurer of buildings and a very significant client for the industry.

Project particulars

In the continuous climate of cost savings and reduction in government departments many relocation exercises of staff are carried out. After years of review it was decided to 'collocate' functions carried out by six government departments in different parts of the UK into one location, for this purpose in Newtown. A 'project sponsor' was appointed as a senior person in AA and he brought together within AA a focus of 'information exchange' in order to form AA's brief for the 'one location' facility. This process took approximately 15 months of consultations with the six departments involved. Once it was decided that this process was finished then the 'project sponsor' entered discussions with another branch of AA and a 'project manager', Mr Drive, was appointed. He had the role of the 'in-house executive' responsible, in the jargon of AA, 'of driving the project to completion'. Mr Drive had experience of the construction industry gained by being a project manager for AA (and its predecessors) over six years and before that by being a contracts manager for a major national contractor. He had not had close experience of construction industry design consultancy.

Main considerations

Drive had been told that 'the Newtown project' was very urgent. The savings in staff costs that would be gained by 'collocation' and by reducing the overall numbers of present staff by being in one place, by reducing inefficient working and so on, were very large and each week's delay in opening would be significant. Very large sums were involved and reputations and future careers for those involved could be in the balance. Time was obviously of prime importance. Because AA had sometimes been criticised for the standards of its design of public buildings it was also anxious to obtain the best possible design for Newtown.

Procurement assessment criteria (PAC)

After a month in his role as Newtown project manager, Mr Drive chaired his first AA meeting and, among other things, stated that he saw procurement of Newtown being achieved by his acting as the 'in-house executive'. Because of his experience and, more importantly, that of AA generally in building procurement (this involved a very considerable annual sum) it was considered appropriate not to seek any further advice on procurement. No principal adviser was required. For the purpose of this case study Drive's initial conclusions are put in the order of the *Thinking about Building* PAC although these were not specifically referred to or arrived at in this way.

- an initial *programme* for procuring the Newtown building facility was estimated to be design of 18 months and construction of 24 months. The estimated construction and consultancy cost was over £200 million. It would be important that once a date for occupation of the building was given this was achieved
- *significant variation* in the building process was not desirable although possible. Once plans had been 'frozen' it was to be hoped that no significant client changes would then occur
- *complexity* of building was involved. A campus of high-quality buildings, requiring innovative design and construction, was needed and major design practices and construction companies would be needed to achieve this
- *the quality level* that was required was clear in one way but not in others. The internal finishes, air-conditioning, standards of sanitary ware and so on would be appropriate to such major office buildings. It was the overall question of external quality, the look of the building, its 'image' about which he was much less sure. His previous experience had not equipped him for judgements in these areas, although he 'knew what he liked' and was concerned that, in his experience, he had found that some architects could become difficult to manage when challenged over their design and needed clear, certain control. If responsibility for managing the design process could be placed with another organisation(s), outside of the AA, then so much the better
- *price certainty* was most important. He had known a number of government projects that had exceeded budget and had seen others that had been referred to the Public Accounts Committee. As project

manager he was concerned to manage this project strictly within the budget already given to him and not to be associated with a project that 'went wrong' in any way

- naturally *price competition* was important for a public agency. Competition at as many stages as possible would demonstrate this and that good value for money had been obtained
- *management and accountability* of the project was not seen by Drive as that much of a problem. The AA had procedures that it had developed over a long time to control all manner of projects. He was under policy instructions to place as much risk as possible outside of the AA and to obtain clear lines of accountability. He had been taught to keep things as simple as possible. He could 'manage' a number of consultants and contractors if this became necessary
- *risk avoidance/allocation* was important to the AA. As a principle as much risk as possible would be placed outside AA.

In summary Mr Drive stated his view of the PAC as:

1. Programme – critical, once a date for a finished building had been given.
2. Price – competition was very important.
3. Variation – should not occur if the AA had done its work properly but in his experience 'project sponsors' sometimes had pressures that required changes, sometimes relatively late in a project. Facility for accommodating change should not be too high in priorities.
4. Quality – the design and finished project was most important.
5. Risk allocation – most important, principally because the Newtown buildings would receive considerable public and construction industry attention.

Procurement arrangement options (PAO)

From the PAC review the following procurement arrangement options (PAO) were established:

(a) Management contracting/construction management
The size of the project, over £200 million, would be suitable for, perhaps, construction management. However the lack of price certainty was a major disadvantage in 'management procurement'. Overall too many risks would be left with AA and this would be a major disadvantage. Drive would need a great amount of argument to allow him to see 'management' chosen for procurement.

(b) Traditional
Design was a very important factor for Newtown. An architect with ability and a significant design team would be required. Drive was unsure how division of responsibility between designers and contractors would meet his accountability standards. Too much division even between consultancies and then between design and construction organisations was not desirable.

(c) Design and build/develop and construct
This was the remaining distinct possibility, provided that the design and quality required could be obtained by that procurement route. This route was one AA had used, almost to the exclusion of other methods, and there was a predisposition towards it.

Drive listened to submissions on procurement and then proposed to his superiors that 'develop and construct' should be adopted. After some weeks of consideration AA superiors confirmed to Drive that his proposal for procurement by some form of 'develop and construct' was accepted. AA decided to proceed within the following framework:

- to invite design and construction in two separate commissions that would be given to different organisations
- first, to appoint one organisation to be responsible for management and provision of all further briefing, design and cost control services in order to produce a scheme for Newtown that would go up to 'scheme design' so that all major elements of the specification were fixed to the satisfaction of AA
- second, to appoint a construction company to develop any design still required to be done and then to construct and complete the facility
- the first organisation would carry on with its commission to monitor the performance of the second organisation throughout the construction period, particularly to ensure that any design developed by the second organisation was acceptable to AA and in accordance with the commission and specification of the first organisation.

Reasons for the procurement choice

The choice of PAO was between traditional and develop and construct. Management was not appropriate because, whilst quality of design and finish could be obtained by that route, there would be insufficient certainty of final price with any form of 'management' and too much risk left with AA.

It was felt that much of the project risk should be placed within as few organisations as possible and that the traditional route would not achieve this because of its separation of design and construction. If faults in the finished building occurred it had generally been found from experience that determining responsibility for fault was not easy.

Postscript

The assembling, shortlisting, tendering and selection of organisations took approximately a further nine months. Tenders were invited for a management/design/cost control organisation to carry out stage 1 of procuring Newtown. The stage 1 commission was to carry out detailed design up to around RIBA Plan of Work, scheme design, stage D. In order to tender for this commission a considerable amount of conceptual design, design drawings, cost estimates, risk analyses and so on were required to be produced over a two- or three-month period in order to show AA assessors the capability of the organisation, if it were to be selected. A fee submission was also part of the selection process.

Over 100 organisations offered their services and a reduced list of around 25 companies was assembled. AA insisted on 'one-point responsibility' and this meant that the consultancy part of the industry had to form a number of 'consortia' in order to tender for the stage 1 commission.

After several changes in the requirements for tendering and other delays, tenders from six companies were invited. Then two were asked to develop their proposals further before one was eventually selected. Industry journals suggested that consultants may have spent well over £750 000 on the selection process and in tendering for stage 1. One unsuccessful consortium of companies in the final list was reported to have spent over £150 000 on its submission for stage 1.

Once selection of the successful stage 1 tenderer was made then the successful consortium carried out development of AA's brief, detailed design to RIBA stage D and associated documentation to allow construction tenders to be obtained. This process took around 12 months.

Many construction companies had gone through their pre-qualification stages and generally had to form 'joint ventures' between construction and environmental engineering companies in order to offer 'one-point responsibility.' A shortlist of companies then tendered for the develop and construct contract and a selection was made. Construction was intended to take around 24 months but delays occurred for a number of reasons, including disputes over how much design was necessary by the stage 2 contractor and how long approval by the stage 1 organisation of stage 2's development of design should have taken (and how long it took). Disputes took a long time in the resolution and the procurement route was not successful in that respect. Quality of product was good but quality of price certainty was not trouble-free.

The overall time from the decision to proceed down a develop and construct route until project completion was around 50 months. For a building of this size and quality this is reasonably fast and the project should be considered a qualified success in its procurement route, apart from disputes of some significance over development of design and associated costs by the stage 2 contractor.

Case study 10: Shopping centre redevelopment

Client

Central Plc was a general property company that specialised principally in the sale, purchase and maintenance of property. It had generally been cautious of taking part in the development process, preferring to leave the risk of that to others. However it had seen companies with apparently little experience succeed in development. It was also aware that it was becoming susceptible to possible takeover bids and it felt that by taking on larger developments it would improve its profile and guard against takeover.

Project particulars

Central had a number of properties in shopping centres in some of the best-known towns in the UK. One of them was in Downtown. The site was part-owned by Central and they believed they could purchase some adjoining properties to make a comprehensive redevelopment. Central was cautious, as this was a departure for them into an enterprise that would take several years to accomplish.

Main considerations

One of Central's development directors brought together key staff responsible for the project and reviewed the position. The main points seemed to be:

- work could not commence until all property on the site had been acquired
- planning permission was expected to be a formality
- quality of design was to be good but not prestige
- the development appraisal suggested that parts of the scheme were marginal and price certainty before starting construction was preferable
- funding would be provided largely from Central's sources and

a fast programme was not a pressure, at least not because of external loan interest.

Procurement assessment criteria (PAC)

The development director agreed with his team that the following PAC were important:

- *programme completion* was not critical. Unless a downturn in the market caused the project to stop there seemed to be no particular haste
- *significant variation* in the design after commencement should not be required. Flexibility in requirements for a major store tenant may become necessary
- *complexity* in the construction was unlikely. Modern appearance and attractiveness for the development in the townscape was desirable with probably traditional materials preferred
- *the quality level* was to be good
- *price certainty* was important. Knowledge of the construction cost before starting work provided a comforting feeling to a corporate client not used to taking risk
- *price competition* was important. No undue urgency over a programme for design and construction was apparent and competitive tendering was a usual procurement process
- *management and accountability* were to remain with Central
- *risk avoidance/allocation* was not considered a factor of much concern.

In summary the PAC review confirmed priorities as:

1. Price certainty – confirmation of major costs before construction started.
2. Quality of design – good but not expensive.
3. Price competition – within limits it was always necessary to demonstrate this.
4. Programme – completion when contracted to be so.

Procurement arrangement options (PAO)

From the PAC review the following procurement arrangement options (PAO) were established:

(a) Design and build, develop and construct
Central had no experience of this procurement method and did not

understand its benefits. They thought it would not suit their way of working with an architect.

(b) Design and manage
Central briefly considered this method.

(c) Management contracting/construction management
No programme pressures, complexity of construction or phasing were present to make a management route appropriate.

(d) Traditional
Central had, on the relatively few occasions that they had built new construction, experience of this method and were familiar with the concept.

Central therefore decided to make the following appointments:

- an architect to obtain planning permission and prepare an overall scheme
- a quantity surveyor, services and structural consultants.

Reasons for the procurement choice

The choice of a traditional route was correct at least for the early part of the scheme. No programme constraint, construction management or construction advice was required and priorities of price certainty and price competition were the relevant PAC.

The scheme proceeded into planning stage and then into working drawings. A major tenant was in prospect for a significant part of the scheme. A year's work was carried out with planning and purchases of parts of the site proceeding in parallel.

Then the local community society showed interest in part of the site that had not yet been acquired by Central. The Downtown society considered that the replacement of some buildings was not necessary or desirable except to the development aims of Central. At the same time voluntary purchase of a number of properties did not proceed according to programme.

Central were unsure how to proceed. They had commissioned detailed design for major parts of the scheme and production of a bill of quantities was under way.

They again reviewed the development with their design team:

- two-thirds of the scheme remained apparently firm and substantially designed, including the major store

- another shopping redevelopment was being planned in Downtown. Central were concerned not to have their scheme completed after the rival one
- PAO were reviewed and a management route considered as an alternative. The major part of design was already completed and the apparent uncertainty of the management system was not thought favourable
- a traditional method was confirmed as the procurement route but the pace was to be quickened and a tender sought, using firm bills as soon as possible. The parts of the scheme that were in some doubt would be covered by provisional sums and quantities.

The design team accepted these decisions, quickened their pace and tenders were sought.

Postscript

Competitive tenders using the traditional, sequential route were obtained. Central's concern with the overall programme of design and construction had resulted in the design team being asked to complete its tender documentation quickly.

The contractor appointed was given a fairly constrained and tight period for construction. During the early parts of the contract bad ground conditions, delayed issue of information, instructions issued to delay parts of the work when principal tenants changed and/or required amendments to work already constructed, conspired to make the contractor request extensions of time and also claim for disturbance to his planned, regular progress.

Central were unused to responding to a design team and contractors when planned progress was not being achieved. Programme and price increases caused Central, their consultants and contractors an unhappy, prolonged experience with the development eventually being completed over nine months behind programme. In shopping developments, often geared to seasonal completion, this is not satisfactory.

Case study 11: Clinic

Client

Selfcare was a small private medical organisation that operated two clinics and one medium-sized private hospital. It was owned by a number of non-UK doctors, principally by three consultants. A good standard of healthcare was provided by Selfcare, at premium rates. If required it had access to loan finance that was provided by a foreign bank. Selfcare had acquired very little knowledge of the UK building industry but as an objective Selfcare was concerned to minimise risks. It thought that perhaps this might be achieved by commissioning projects stage by stage wherever possible, so having the ability to stop a project before construction started if the perceived risks appeared to outweigh the likely benefits.

The two clinics operated by Selfcare had been purchased by them and they had so far no direct experience of the design, construction and establishment of a clinic of the size that they now intended. The hospital that Selfcare operated was in London and had been a joint venture with another private medical organisation that had provided the briefing and procurement initiatives. The other organisation had played the major part of the joint venture and it had made the decisions on procurement, appointment of consultants, contractors and so on. Selfcare had noticed throughout the construction of the London hospital that the process of design and construction seemed to them to be quite unpredictable. The actual time taken in the design and construction of the London project and its final price were both considerably beyond initial indications. To those familiar with construction there were good reasons for this (for instance, considerable changes occurred in the brief and many variations were necessary throughout the design period and also throughout the construction period). However to Selfcare these did not seem sufficient as reasons to extend the programme and to increase the final price so significantly.

Project particulars

Selfcare now wished to build a private clinic to provide residential care and minor/medium operation facilities in Provincetown. The approximate construction cost at that stage had been suggested to Selfcare as £15 million. A healthcare marketing consultant had provided survey research on the income groups within the catchment area that would be served by the clinic and had made a case establishing the need for such a facility. Feasibility studies on the design and further development appraisals would need to be progressed.

Main considerations

The three doctors who were the principal private shareholders in Selfcare were concerned to make the Provincetown clinic a notable development, not only in private healthcare but also as a facility that would establish the new image they wanted to create for Selfcare. In their view the clinic must look impressive and even prestigious. The foreign bank would have to ratify all major decisions on procurement and expenditure but the briefing and establishment of medical requirements would naturally come from the Selfcare doctors.

Early financial appraisals of the clinic had been provided, unasked for, by the healthcare marketing consultant commissioned by Selfcare and the construction value of approximately £15 million given by them was encouraging. Income predictions showed that a significant surplus over expenditure should occur, such that it appeared that if any loan finance were required, it could be repaid within the first five or six years of the clinic's operation. The clinic would then be owned by Selfcare with no debt liability.

The three doctors then held discussions with their bank who understood that Selfcare could only finance approximately 40 per cent of the project themselves. The bank therefore agreed to provide a loan of 60 per cent of the final project cost for a period of six years after construction completion. During the design and construction phases the bank would provide all interim finance to Selfcare.

One of the three doctors and a senior executive of their bank became the in-house project executives for the Provincetown clinic. The two-man group did not formally recognise this but in reality they provided any executive action required. The doctor had views on the operation of a clinic, and therefore on its design, and the banker had views on financial matters. The doctor, as a foreign doctor, had travelled extensively and he had seen architecture from many cultures and countries. He had views on design, that is, principally on the appearance of buildings. The bank executive was also a widely travelled man who appreciated financial and some contractual and general procurement matters. He did not have a knowledge of UK construction and consultancy matters.

Procurement assessment criteria (PAC): first review

The doctor and the banker (for those were their prime roles) met to discuss the development of the Provincetown project. They already understood they would need to form a brief for their requirements but they knew little more. Through 'connections' in the bank and enquiries of the healthcare marketing consultants they obtained the names of a number of architects with experience in designing hospitals and clinics. They interviewed three or four practices and decided to appoint an architect. No other appointments were made at that stage.

The architect at his first meeting with Selfcare reviewed the PAC. The review was not systematically done, far from it, but for clarity of presentation in this case study the following PAC were established:

- *programme completion* as early as possible was preferable. Loan finance would mean interest payments. Also the marketing consultants had shown very favourable income projections and therefore early completion would be financially attractive. An overall period of perhaps 36 months for the design and then the construction of the clinic was talked about. No options over procurement alternatives were discussed. The architect talked broadly of 'traditional procedures' but nothing specific was explained or presumably understood by this term. The architect had worked exclusively for the National Health sector of the market and felt that, once a brief had been settled, his suggestions on programme were a good guideline
- *significant variation* was not expected. Selfcare did not understand the process of briefing and design that would be required for a clinic but they stated that 'no variations would be required once Selfcare had agreed their brief'
- *complexity* was not apparently involved. The architect had designed several hospitals and health centres for National Health authorities and was confident he could design the health facilities for a private clinic
- *the quality level* was to be high. The Selfcare doctor talked enthusiastically about clinics and hospitals that he had seen in the USA and in the Middle East. He wished to emulate this quality in 'his clinic'
- *price certainty* would of course be important. Selfcare had decided to proceed on the basis of the healthcare marketing consultant's projections, including outline costs. Selfcare asked the architect to comment on the budget proposed for the clinic as soon as possible. The architect in turn advised that a quantity surveyor should be appointed to provide cost consultancy advice and he agreed to propose one or two quantity surveying consultancies
- *price competition* was discussed. Selfcare naturally believed in principle that competition was preferable. Nothing more was discussed on how this should be achieved
- *management and accountability* were discussed and in principle Selfcare wished to pass risks to their designers and constructors
- *risk avoidance/allocation* was important. Selfcare wished to limit their risk and to be told of any risks that they were accepting.

After the first PAC review no conclusions were confirmed nor were any priorities established, even in outline. (In retrospect this was a critical time when the inexperience of the client and the lack of clarity of the architect combined to allow a drift to occur in the project.)

Following the first PAC review four months passed:

1. The architect obtained agreement to his commencing to prepare a brief for the clinic and the appointment of a quantity surveyor.
2. The Selfcare doctor attempted to take part in the briefing process. He was not readily available, among his other duties, to agree the brief and the process took approximately four months.
3. Meanwhile a quantity surveyor had been appointed and he confirmed that the budget given by the healthcare marketing consultants was of the correct 'order of cost' for a clinic. He qualified his confirmation by reference to it 'being in relation to the information made available to him for his estimate and without a programme having been established'. The budget was confirmed at current cost and no indication or estimate of future price inflation was given.
4. An overall programme for design and construction was not kept in mind and the brief and design period took much longer than expected.

Selfcare was concerned at the slow progress in commencing detailed design work and was anxious also to start construction. Selfcare reminded the architect that there was a market for the new clinic and that Selfcare wished to satisfy that market by constructing and operating the clinic. The principal doctor from Selfcare stated that 'overseas we would have started construction by now. The architect would have prepared outline plans and the contractor would have started work.'

Procurement assessment criteria (PAC): second review

Approximately six months after the decision by Selfcare and their bankers to appoint an architect a further PAC review took place. The doctor, the banker, the architect and the quantity surveyor agreed:

1. Programme completion – apparently this was becoming more important, as a priority, 'namely to open the clinic'.
2. Complexity – no further comment.
3. Variation – this was not seen as a problem.
4. Quality – this remained an ill-explored area. This misunderstanding was not appreciated either by Selfcare or by the architect and the quantity surveyor.
5. Price certainty – this remained important.
6. Price competition – this remained an objective.
7. Management – this remained ill-defined.
8. Accountability – this remained unclear.
9. Risk avoidance – risks were not explored or further defined.

Procurement arrangement options (PAO)

From the second PAC review the following procurement arrangement options (PAO) were established:

(a) Design and build
This was accepted as being inappropriate by Selfcare, by their bankers and by the design team. The brief was hopefully now almost settled and the appointment of a contractor to develop the design was not felt appropriate. It would have meant that many design decision would be taken by a design and build contractor and this was not wanted by the architect – 'quality would not be achieved that way', in his opinion.

(b) Traditional
If the overall programme would allow, then this was the procurement path favoured by the architect and the quantity surveyor. They were fully conversant with it and preferred the traditional checks in 'the plan of work'.

(c) Management contracting
This procurement method was explained to Selfcare and the prospect of a shorter overall programme appealed to them. If they could save financing costs of perhaps six to eight months and have an operational clinic that much earlier surely there was much to be said for the management route.

It was apparent that although PAO were reviewed no firm commitment to one of them was made.

Outline planning permission had by now been obtained by the architect for Selfcare and it was agreed that the architect should obtain detailed permission. The quantity surveyor prepared revised estimates of cost for the clinic including forecasts of inflation. Inflation in tender prices had been relatively low for several years and at that stage general indications were that it would continue that way.

Selfcare, having agreed the brief, began to introduce the possibility of some changes. They wanted to introduce a major operating theatre and, to keep flexible for a change they perceived in healthcare, they wished to expand provisions for alternative medicine facilities. They agreed to some increases in the budget to allow for these changes whilst continuing to question the tendency in each financial report for the overall costs to increase (unjustifiably in their eyes). Selfcare also received around this time information that a neighbouring town had also received a planning application for a clinic. Understandably Selfcare became increasingly concerned to 'have work started'. Many reviews of programme, overall budgets, quality and so on took place whilst the architect and an expanded design team were commissioned to proceed as quickly as possible. The design changes that had begun to be introduced caused some delay and disruption to the usual flow of design.

Selfcare continued to push for an improvement in the expected date for completion of the clinic. Briefing and design had now taken nearly twelve months and the completion of construction was, by the traditional route, said to be at least 30 months away.

The architect and quantity surveyor, under some pressure, proposed that the alternative of a management route was perhaps worth exploring. They cautioned, in retrospect it appeared not very strongly, that price certainty and risk avoidance were not so achievable by a management route.

However it was agreed to ask a number of management contractors to submit proposals for the management of the construction contract. In this way, it was said, 'we will all gain the benefit of a contractor's opinion on programme, on buildability and so on. Perhaps we are all being too pessimistic about programme and also about cost. A contractor's view could help us here.' Some of the management contractors suggested that completion could be achieved in a very acceptable (to Selfcare at least) time.

A management contractor was appointed and the design programme was adapted to allow 'packages' of work to be let to works contractors. Selfcare continued to introduce a number of changes in their requirements (for instance air-conditioning to parts of the residential accommodation), the rate of inflation in tender prices increased over the 27 months of construction and the budget that had been set at contract stage was considerably exceeded.

At several stages within the first nine months of construction Selfcare considered very carefully stopping the construction contract. Their bankers assumed effective control of the project and required Selfcare to contain their involvement and to stop the introduction of any further changes in requirements. The action of the bank was sufficient to establish a climate that allowed the design and construction process to be satisfactorily completed, some nine months later than the management contract period. Quality of the finished product was good.

Postscript

There was no in-house executive at Selfcare who was capable of performing the role required for such a project. (This role has been described in Chapter 2.) The combination of a doctor and a banker was the best available solution but it was not adequate to the task. The healthcare marketing consultant and the architect were not appointed as, nor did they seek a role as, external principal advisers. The architect performed a reasonably competent role as an architect but he did not successfully explain the full constraints and opportunities of the procurement options on the several occasions when such an explanation was essential. The architect and quantity surveyor were pressed into adopting a management route and failed to emphasise the considerable risks that Selfcare would have to accept in taking that procurement route. The project was completed but not to a programme or to a price that Selfcare wanted. Had they known in the first place the full price and delay in having an operational clinic they would not have proceeded with the project.

Selfcare had to borrow more than they had intended, over a longer development period than they had planned. They were very satisfied with the product they ultimately received and, because their initial income projections were also easily surpassed once the clinic opened, due to inflation and to an increased demand for use of the clinic, they eventually operated financially in advance of their expectations. The client, consultants and contractors were not satisfied with the procurement process and did not wish readily to repeat the experience.

Case study 12: Shopping centre renovation

Client

Shopcentres was a property company specialising in the acquisition and development of small shopping units. They believed that small groups of shops could be developed, often by renovation, and they looked for opportunities for acquisition where they were not likely to be in competition with major property and development companies.

Project particulars

Monksgate was an open square with shopping units on four sides and three access points into the square. Along the access points there were shopping units, mixed with some residences, in open streets. Shopcentres acquired the freehold of the open square and some of the streets and wished to improve the shopping facilities by covering the square, covering two of the 'streets', and renovating/altering a number of the shop units. At that time two other shopping areas in the town were known to be considering similar changes.

Main considerations

Shopcentres wished to carry out their improvements quickly, to a known price. They wanted good quality of product.

Procurement assessment criteria (PAC)

A Shopcentres' in-house executive was appointed to progress the project and he met his internal property and construction advisers.

A review of the PAC suggested:

- *programme completion* was desirable in order to suit trading of the renovated 'centre' before Christmas, in 15 months' time
- *significant variation* was not considered likely
- *complexity* was not present. No complex construction or environmental services were likely to be involved
- *the quality level* was to be of good materials and workmanship
- *price certainty* was important but within reasonable limits
- *price competition* was required unless other PAC prevented it
- *management and accountability* were important. Shopcentres preferred to delegate controlled responsibility to consultants and to contractors
- *risk avoidance/allocation* was required to be known by Shopcentres.

In summary the PAC review established priorities as:

1. Programme – an ideal programme for completion had been set.
2. Price – competition was desirable.
3. Variation and complexity – these were not contemplated to be significant.
4. Quality – this was required.
5. Risk allocation – required to be clearly established.

Procurement arrangements options (PAO)

From the PAC review the following procurement arrangement options (PAO) were established:

(a) Management contracting
The size of the project, about £2 million, was not appropriate for management contracting, nor was the relative simplicity of construction. It was accepted that competition and programme completion were obtainable by management contracting.

(b) Traditional
Provided the completion programme could be met by this route traditional procurement offered advantages in price certainty and in competition, provided variations were avoided.

(c) Design and build, develop and construct
No advantages were found for this route. Different design options were not wanted from a number of contractors and develop and construct did not seem to give an advantage. The time that the architect would have to spend in preparing comprehensive client's requirements, and then evaluating design and build tenders, could equally well be spent in preparing his design and seeking a traditional tender.

Shopcentres' executive made the following decisions:

- the appointment of an architect and design team to obtain planning permission and produce a design from which competitive tenders could be obtained
- that the brief to the design team required an overall programme of design and then completion of construction within 18 months.

At a subsequent meeting Shopcentres' executive reviewed the consultant's preliminary designs and their procurement proposals, within the considerations he had previously given. It was agreed that:

- the programme required an early contractor appointment, before all design would be completed. A construction contract of approximately twelve months was probably required
- an accelerated traditional route would be used. Competitive tenders would be invited, based on such approximate quantities as were available after three months of design. Provisional sums would also be provided for lesser elements of the works.

Reasons for the procurement choice

The choice from the PAO was between either an accelerated traditional or a management contracting route. Variations of design and build were not appropriate because of the time needed for even elementary design to be settled before develop and build tenders could be invited. Evaluation of different design and build proposals was not welcomed. Obtaining a contractor on partial design information and approximate quantities was thought to give a satisfactory, fast method of early contractor appointment to suit a short programme. Price certainty at the start of construction could not be guaranteed this way but earlier programme completion was more important and hopefully would be achieved by an earlier contractor appointment and start of construction.

Postscript

Subsequently Monksgate was converted into enclosed shopping facilities according to the programme. Price exceeded the initial budget and also the contract sum provisions, partly because some variations were introduced. The procurement choice was successful.

Case study 13: Golf country club

Client

Bogey Leisure was a general leisure company that provided private health clubs, hotels with health, club squash and badminton facilities. As many marketeers had stated during the 1980s, leisure was a growth industry and could not easily be confined and defined by boundaries. Bogey wished to expand into golf/leisure facilities and had their first project in prospect.

Bogey did not have extensive experience of the construction industry.

Project particulars

Bogey Leisure had acquired over 200 acres of land in Derbyshire to construct a golf course and associated residential and country club facilities. The golf course design and construction, and residential developments within and around the golf courses, were to be developed by a separate procurement route but a golf club/country club was required at the start of the leisure development. It was to cater for a membership of 200 and provide a country club/leisure facility for members and visitors of, at peak, 400 with health club, swimming, badminton and squash provisions.

Main considerations

Bogey wanted a facility to be erected as the first phase of the overall development, alongside the construction of the first 18 holes of the golf courses. They delegated one of their in-house executives to propose how this should be obtained. He decided his priorities as follows:

- he wanted construction to be provided as soon as possible and he wanted to know its price before he started

- he had an idea of quality standards but only by reference to other similar facilities he had used and/or stayed at.

Procurement assessment criteria (PAC)

The Bogey executive sought advice from a project manager whom Bogey Leisure had used for coordination of renovation works to hotels within the Bogey enterprise. Together they reviewed requirements and priorities, not as set down below, but in a less structured way. For clarity the salient points have been listed against the usual PAC.

A review of PAC suggested:

- *programme completion* was very important to establish a presence and an income for the Derbyshire project as soon as possible
- *significant variation* was not considered likely. It did not usually occur in Bogey's operational style and it did not seem to be a PAC
- *complexity* in the building was not contemplated. Although it would have multi-use facilities, including swimming, catering, and some residential, it was not considered to be complex
- *the quality level* was to be good in order to attract business executives and managerial members
- *price certainty* was wanted, as in the majority of Bogey's operations
- *price competition* was considered very necessary
- *management and accountability* should ideally be placed with one organisation to design and construct
- *risk avoidance/allocation* was important to Bogey and should be placed clearly with another organisation where practicable.

In summary the PAC confirmed priorities as:

1. Programme – a club house was required as soon as possible.
2. Price – this must be known before an order to start was given.
3. Product – a good quality was required.
4. Competition – this was required for a project of around £5 million in value.

Procurement arrangement options (PAO)

From the PAC review the following procurement arrangement options (PAO) were established:

(a) Traditional
This was considered to take too long and to lack control, with design and construction being separated.

(b) Management
This was not easily understood and, because of the probable involvement of a project manager advising Bogey, was not considered to be appropriate.

(c) Design and build
This route was considered to be very appropriate because it placed responsibility with one organisation.

Bogey Leisure then decided to make the following appointment:

- a project manager/client's representative. This was the same organisation that had assisted the Bogey in-house executive in a review of PAC. The project manager commenced implementation of procurement along a design and build route.

Reasons for the procurement choice

Bogey Leisure had little in-house resources or expertise to devote to managing the procurement of a country club facility. They decided therefore, wrongly as it turned out, that they could, once having obtained a project manager, through him then obtain a design and build contractor to provide the facility. The skills and knowledge of the project manager and of the contractor appointed were not capable of providing Bogey's requirements.

Postscript

The external project manager attempted to obtain a brief from Bogey Leisure, in order to prepare a 'client's requirements' document that could then be used to obtain tenders from appropriate design and build contractors. Bogey however were not clear about their requirements and eventually, in order to seek tenders from design and build contractors, it was agreed that the 'client's requirements' documents should state only spatial and functional requirements and contractors would then make proposals in return for providing these requirements.

Tenders were eventually received from four design and build tenderers and, not surprisingly, were found to offer quite different solutions in price, in the quality to be provided and in the space/functional proposals. Analysis of the tenders was carried out by a firm of design consultants, on behalf of Bogey Leisure. A lengthy period was spent in comparing quite different tenders and in trying to obtain a clear match of 'client's requirements' and 'contractor's proposals'.

The tender that was eventually accepted still contained a number of provisional requirements and proposals and the standards of quality being offered were not clear. The contractor had to obtain planning permission, which he did after some modifications were made to his design. Once some of the finer details of his proposals became evident to Bogey Leisure they insisted on changes and price rises and programme delays occurred.

The country club was finished considerably later than intended and protracted negotiations were necessary over price extras and quality standards. Consultant advisers were employed by both the client and the contractor to aid in the post-completion negotiations.

Inadequate internal Bogey resources were made available to the project at the appropriate times and an unclear, ill-developed brief was used to obtain tenders from design and build tenderers. Design and build can be an excellent procurement method but it does require skilled drafting of the 'employer's or client's requirements'. Ambiguities at contract stage were not resolved in this case until during the course of construction and this resulted in client and contractor disappointment and disagreement.

Case study 14: Rebuilding of church destroyed by fire

Client

St Dunstan's Church was extensively damaged by fire in late autumn and, to all intents, it had been destroyed. In the late 1840s it had been designed to seat 700 people. Now congregations of around 250 at most were more usual, with under 200 more the norm. The vicar met with his parochial church council (PCC) within a week of the fire and began the process of considering building renewal. St Dunstan's had a sister church in the same neighbourhood and was therefore able, at some inconvenience, to serve its congregation by holding joint services until St Dunstan's was rebuilt.

Project particulars

The PCC of St Dunstan's had within its members a bank manager, a general practice surveyor, a Lloyds underwriter, a quantity surveyor and a project manager who had retired early from a major oil company. The PCC decided at its first meeting to form 'a building committee', chaired by the ex-oil company project manager, Mr Shells. Among other things he decided to ask the church parishioners about rebuilding and also the diocese about any matters they may wish to have considered. He also knew that it was necessary to contact the insurance company that had insured the building for loss arising from damage by fire, to find out how much money they would pay for any rebuilding. At around the end of January of the following year Mr Shells called together the PCC building committee to review rebuilding plans.

Main considerations

The diocese authorities indicated that they would like rebuilding to be in a more modern style, rather than to rebuild the Victorian building exactly as it had been before it was destroyed by fire. A programme for carrying out the work was not too critical as it was understood that the tender market was very competitive and that 'builders were crying out for work'. Also as church activities and services could be held at the sister church for the foreseeable future it seemed more important to rebuild after full consideration of all matters. A number of local builders and local architects had already written to St Dunstan's and telephoned to say that they would be very interested in carrying out the church's proposals for rebuilding.

It was agreed that the following week the PCC committee should consider a 'starter paper' that would be put to them by Mr Shells.

Procurement assessment criteria (PAC)

After a week of reflection Mr Shells, acting as the church's 'in-house executive', suggested that:

- he did not consider that other expert advice, for instance from *a principal adviser*, was required beyond advice that would continue to come from members of the PCC. Although the PCC had not previously been responsible for building procurement on the scale proposed for the church rebuilding Mr Shells considered that he could act, if appointed, as the 'in-house executive' on a voluntary basis. He would accept this as an honour, a duty and a responsibility to the community of St Dunstan's
- a *programme* for rebuilding should take around 18 months. It would be appropriate but not vital if the work could be finished in time for the one hundred and fiftieth anniversary of the building in three years' time
- *significant variation* in the building process was not likely after full plans had been agreed

- *complexity* of rebuilding was likely to be very much of a consideration
- *the quality level* that was required was causing him some concern. The parishioners, at least those that had responded to his questionnaire and to meetings that he had held, were divided fairly equally between those that wanted the past rebuilt and those that wanted to 'go for something more modern', something more suitable for the twenty-first century, not a replica of the past
- *price certainty* was also causing him some concern and he was strongly recommending that the PCC should only go for a 'fixed price' contract
- naturally *price competition* was important because the insurance company would be providing most of the money with the balance coming from diocese funds and possible gifts from a number of the wealthier parishioners. All these sources of funding would want to see competition
- *management and accountability* of the project were not seen by Mr Shells as too much of a problem. He was willing to offer his project management experience to the project and he suggested that the PCC should go for a system of procurement that placed as much of the risks as possible with a contractor. He favoured keeping things as simple as possible but he could 'manage', on behalf of the PCC, a number of consultants and contractors.
- *risk avoidance/allocation* was important to the PCC but provided the risks were known in advance Mr Shells considered that sharing of risk was probably the fairest and most practical solution for the PCC and its consultants/contractors.

In summary Mr Shells stated his view of the PAC as:

1. Programme – was not critical within limits.
2. Price – competition was important.
3. Variation – should not occur if the PCC had done its work properly.
4. Quality – of design and finished product was very important.
5. Risk allocation – was important, principally because the church would not be able to overspend once its aggregate of money from insurance, the diocese and a few parishioners was set.

Procurement arrangement options (PAO)

From the PAC review the following procurement arrangement options (PAO) were established:

route, around 3–6 months may have been saved but at some risks to price certainty. This client was unable to save very much time in rebuilding because at the beginning it was uncertain that it wanted simply to rebuild. Instead it took the opportunity to explore possibilities of rebuilding less of the building than existed at the time of the fire, interviewing a number of architects to test their views and capabilities and then going for a more modern design and a smaller building than that destroyed. The client's initial caution and careful reviews with parishioners and its insurance company were worthwhile in the long run, although two church congregations had to put up with 'sharing' facilities of the sister church for a long time.

General lessons

The overriding lesson is that a project is only as successful as its client.

Now, in 1996, hardly a week goes by without the construction professional and trade press stating 'how important the client is' or 'the construction process is all about clients, and meeting their wishes and aspirations . . . satisfying the client must be the ultimate objective' (this is a quotation from a well-known trade journal discussing the Latham Report in August 1996 – two years on and all that). The statements may appear simple and trite but they cannot be emphasised too much or repeated too often. If a client does not know what he wants, more importantly what he needs, he will almost certainly be dissatisfied with the building, the product, that he eventually receives.

- If a client does not know what he wants he needs very expert advice before he does virtually anything else. Discussion within his own organisation will hopefully clarify some issues. Conversations with respected outsiders, accountants, solicitors, bank managers, maybe principal customers and so on may help.
- Then if he has not experienced building procurement and all its potential vicissitudes, he must talk to and obtain independent advice from a 'principal adviser' (PA).

I have explained, mainly in Chapter 2, but also throughout the book, the essentials of the role, the qualities required to carry out the role and how the advice and counselling of a PA is usually vital. Clients that do not build regularly and those that have not built before should appoint a PA.

- A principal adviser who is knowledgeable of sufficient aspects of the construction industry and also of the client's business may often be quite hard to find. The major professional institutions (RIBA, RICS, ICE) should be able to provide a list of organisations that can advise.

- The PA appointment should be for advice in the PA role only. No extension or conflict of interests should be created by expanding the PA role into project management, design, cost control, construction management or construction. It is as though the PA is a non-executive board member for the project duration – his duties should not be expanded into executive actions.

A client that is fortunate to know his requirements, and to have procured them or similar ones before, should have the experience to proceed along a procurement path with some confidence. His consultants and contractors should benefit from this, provided they also are organised and able to contribute as and when required. The case studies of successful projects have shown this. In Case studies 4, 10, 11 and 13, the inexperience of the client, and to some degree of the consultants, meant that all concerned were, as things turned out, going to have some relatively unhappy, bad experiences.

- If product quality, that is, the level of quality of design and finished product expected, is fairly well-known by a client and his team it should be possible to obtain it. It should be identified and allowed for at the outset of a project and the variables of product programme and product price should then be put into priority order.
- Unfortunate experiences in building procurement very often revolve around vacillations between programme certainty and price uncertainty. A client is often torn between a procurement method that may mean a longer programme but may be more price-certain, and a procurement path that has a shorter programme, perhaps by using more management and other resources, but carries a price uncertainty with it.

Cases: relative success/failure of procurement route adopted

The range of judgment employed on procurement success has been:

- very successful
- successful
- relatively successful
- relatively unsuccessful
- unsuccessful

Case Study No. Project/Client	Procurement path	Relative success/ failure
1. Supermarket; Supermarkets Plc	Traditional accelerated	Very successful
2. Offices, shopping and leisure facilities; Citystyle	Management	Very successful
3. Hotel extensions; Retreat Hotels	Develop and construct	Relatively successful
4. Civic hall and theatre; Sometown Borough Council	Traditional accelerated	Unsuccessful
5. Offices; Growth Developments	Design and build	Successful
6. Factory; Terrestrial Industries	Design and manage	Relatively successful
7. Department store, phased redevelopment; Insurance Plc	Management and traditional	Successful
8. Hospital; regional health authority	Traditional sequential	Successful
9. Prestige building; government agency	Develop and construct	Relatively successful
10. Shopping centre redevelopment; Central Plc	Traditional sequential	Relatively unsuccessful
11. Clinic; Selfcare	Management	Relatively unsuccessful
12. Shopping renovation; Shopcentres	Traditional accelerated	Successful
13. Golf country club; Bogey Leisure	Design and build	Unsuccessful
14. Church rebuilding; St Dunstan's	Traditional	Very successful

Glossary

Accelerated contract A traditional form of procurement where a contractor is appointed on partial design information and a schedule of rates or a bill of approximate quantities from which a contract is negotiated.

ACE The Association of Consulting Engineers.

Actual cost The sum of costs incurred by a contractor in performing a contract. It is used when a contractor is reimbursed using a target cost or cost reimbursement mechanism.

Alliance of Construction Product Suppliers This represents materials and products suppliers on the Construction Industry Board.

BOOT A method of procurement in which one organisation builds, owns, operates and then transfers its ownership to the commissioning authority. In the intervening period it operates the facility under licence.

BPF The British Property Federation.

Client A person who commissions the design and construction of a building. The initiator of a project is commonly referred to as 'the client'.

Client's agent See under 'employer's representative'.

Construction The business or work of building; assembling materials, using labour and plant to form a building.

Construction Clients Forum This body was formed during the Latham process and represents regular and occasional/one-off clients. It is represented on the Construction Industry Board.

Construction Industry Board (CIB) Organisation composed of the Construction Industry Council, the Construction Industry Employers Council, the Constructors Liaison Group, the Construction Clients Forum, the Alliance of Construction Product Suppliers and the Government.

Construction Industry Employers Council This body represents main contractors and is a member of the Construction Industry Board.

Construction management contract This contract is similar to a management contract except that the client is the employer in each 'works construction contract'. A construction manager becomes the client's agent for management of a contract.

Contractor A person who contracts with another. In the context of building construction, a contractor is the builder who contracts with a client or employer.

Contractor's proposals Proposals made by a contractor, within a design and build contract, in response to the employer's requirements, for the design and construction of a building giving materials, methods of working, programme and contract price.

Coordination A balanced and effective interaction of separate actions. An act of producing a complete and integrated solution of design, of construction, or of design and construction.

Cost-based contract A cost-based contract is one in which a client pays a contractor on the basis of the actual cost of the construction work carried out. Cost reimbursable and target contracts are in this category.

Cost reimbursable contract This is a contract in which a contractor is reimbursed the actual costs incurred in carrying out the work. In addition management, overheads and profit are charged on a fee basis.

Design The exercise of those functions that use scientific principles, technical information and imagination to define a project capable of meeting specified requirements with economy and efficiency.

Design and build A generic term for contracts in which a major element of the design is the contractor's responsibility. They may be divided into 'develop and construct', 'package deal', 'turnkey', 'Private Finance Initiative' and 'BOOT' contracts.

Design and manage contract This contract is similar to a management contract but it involves the design as well as the management of a project being executed either by a design and manage contractor or by independent designers, under contract to a contractor.

Develop and construct contract This is a variation of a design and build contract where a single organisation accepts full responsibility for both design and construction, except that in addition to this a client or employer retains an independent designer or scope designer to provide initial advice, perhaps some schematic design, or nearly full design, and/or to exercise some control over the detailed design developed by the contractor.

Employer A term often used in building contracts and contracts for professional services to describe the client who commissions a building.

Employer's representative An agent employed by a client to act on his behalf with limited powers, for instance, within a design and build contract. Also sometimes called an 'employer's agent' or 'client's agent'.

Employer's requirements A statement of the employer's (the client's) requirements in terms of spatial and quality standards when inviting tenders for a design and build contract. The requirements may be little more than a description of accommodation required, or they may be anything up to a full scheme design. The contractor's proposals (see earlier) should complement the employer's requirements.

Fast tracking A method of reducing project time by the overlapping of design with construction.

Form of contract The document that comprises the conditions of contract. See also 'standard form'.

Guaranteed maximum price If the scope of works can be sufficiently defined, some contractors are willing to quote a guaranteed maximum price (GMP). The contract is often cost reimbursable up to this point, beyond which it becomes fixed. It is a variation sometimes introduced into construction management.

ICE The Institution of Civil Engineers.

JCT The Joint Contracts Tribunal. A body responsible for producing and reviewing standard forms of building contract. It currently comprises twelve constituent bodies and has produced standard forms for use by private and public clients who wish to contract for traditional, design and build and management contracting.

Joint venture A joint venture is the combining of the abilities of two or more companies to be jointly and severally bound to accomplish a project, on the basis of sharing profit/loss.

Lump sum contract This is a contract in which a single price is given for the work. Payment is made either on completion of the work or at regular periods or at stages related to key events in construction. 'Traditional' contracts are usually examples of 'lump sum' contracts.

Management In the context of building procurement and contracts, the term means the activity of managing, organising, supervising and securing the carrying out and completion of activities.

Management contract This is a contract in which management is regarded as a separate discipline and responsibility from that of construction. A management contractor is generally precluded from undertaking any of the construction work. Construction contractors (known also as package, trade or works contractors) contract with a management contractor, who is therefore their client or employer.

Measure and value contract This is a contract based on bills of quantities or schedules of rates that incorporates the principle of payment by the measurement of completed work valued at tendered or subsequently negotiated rates.

Negotiation Discussion, written or oral, between two or more parties on different sides with an aim to reach a common agreement.

NJCC The National Joint Consultative Committee for Building was a consultative organisation comprising six constituent bodies. It was responsible for the encouragement of good practice in the building industry by setting standards for tendering procedures and reviewing, through its Good Practice Panel, the use of standard forms of contract and the abuse/undue amendment of contract terms and recognised, fair trading procedures. It was disbanded in June 1996 and the work of ensuring good practice in the industry has been taken on by the Construction Industry Board.

Package deal   An alternative name for design and build.

Partnering   An arrangement whereby people are encouraged to work more efficiently together including shared problem resolution, continuous improvement, continuity of work, fast construction, completion on time, lower legal costs and improved profits.

Plan of work   A plan or order of working for a design team. The RIBA produced its *Plan of Work* in 1964 to define a systematic plan of sequential operations from project inception through to project completion and feedback.

Price-based contract   This is a contract in which payment for work is based on a price quoted by a contractor for work. The price need not necessarily be related to its cost. Additional work is valued on the basis of contract prices or prices derived from them. Lump sum and measure and value contracts are in the category of price-based contracts.

Principal adviser   A person or organisation experienced and knowledgeable in the building industry that can advise a client, impartially, on the need to build and how to go about building procurement.

Private Finance Initiative   The requirement of HM Treasury to have capital projects examined for funding by private finance before allowing them to proceed.

Probability   The likelihood that an event will occur. Insurance premiums are governed by the cost of an event and its consequences and by the probability that it will occur. These considerations taken together determine the risk.

Procurement arrangement options (PAO)   The options for procurement, described in Chapter 4, for obtaining design and construction services.

Procurement assessment criteria (PAC)   The criteria, described in Chapter 5, that need to be considered in determining a client's priorities in building procurement.

Project management   The overall planning, control and coordination of a project from inception to completion aimed at meeting a client's requirements and ensuring completion on time, within cost and to required quality standards.

Project manager   A client's representative in the management of a project, generally from inception to commissioning. He may be a client employee, an external consultant, or a contractor with a brief to act in this capacity.

Remeasurement contract   See under 'measure and value contract'.

Responsibility   The acceptance of accountability for one's goods or services, that possibly leads to acceptance of liability for defects in those goods or services and/or adverse consequences, but may not necessarily do so.

RIBA The Royal Institute of British Architects.

RICS The Royal Institution of Chartered Surveyors.

Risk An unforeseen or uncertain event associated with a probability for potential loss or gain. Hence whatever is to be insured against is usually referred to in an insurance policy as 'the risk'.

Scope designer An independent designer retained by a client to provide initial advice and/or to exercise some control over the detailed design in a develop and construct contract.

Standard form This is a form of contract produced by a recognised body, for instance the Joint Contracts Tribunal. The term also refers to forms in common use.

Target contract A target contract is a cost reimbursable contract based on a predetermined cost that is adjusted for changes in the works. A contractor's actual costs are monitored and any difference between actual cost and the target cost is usually shared between a client and a contractor in a specified way. There is usually a separate fee for management, overheads and profit.

Turnkey A variation of design and build procurement.

Type of contract The type of contract determines the contractual relationship between the building owner (or his agent in a management contract) and the construction contractor. It usually defines the payment mechanism and the allocation of responsibility and liability between the parties.

Two-stage tendering A variation of tendering (traditional or design and build) where a contractor or contractors are selected on a basis of partial design or tender information, from which negotiations leading to a contract are carried out, once further information has become available.

Works contractor A contractor that carries out construction work under a management contract or construction management arrangement.

Index